The Hand Behind Unmanned

BRIDGING THE GAP

Series Editors
James Goldgeier
Bruce Jentleson
Steven Weber

The Hand Behind Unmanned

Origins of the US Autonomous Military Arsenal

JACQUELYN SCHNEIDER
JULIA MACDONALD

OXFORD
UNIVERSITY PRESS

Oxford University Press is a department of the University of Oxford.
It furthers the University's objective of excellence in research, scholarship,
and education by publishing worldwide. Oxford is a registered trade mark of
Oxford University Press in the UK and in certain other countries.

Published in the United States of America by Oxford University Press
198 Madison Avenue, New York, NY 10016, United States of America.

© Oxford University Press 2025

Library of Congress Cataloging-in-Publication Data
Names: Schneider, Jacquelyn, author. | Macdonald, Julia, author.
Title: The hand behind unmanned : origins of the US autonomous military
arsenal / Jacquelyn Schneider, Stanford University, Julia Macdonald, University of Denver.
Description: New York, NY : Oxford University Press, [2025] |
Series: Bridging the gap series |
Includes bibliographical references and index. |
Identifiers: LCCN 2024041624 (print) | LCCN 2024041625 (ebook) |
ISBN 9780190064389 (hardback) | ISBN 9780190064402 (epub) |
ISBN 9780190064419 (ebook)
Subjects: LCSH: Military engineering—United States. | Drone aircraft—United States.
Military weapons—Technological innovations—United States.
Classification: LCC UG23 .S37 2025 (print) | LCC UG23 (ebook) |
DDC 355.8/2–dc23/eng/20241028
LC record available at https://lccn.loc.gov/2024041624
LC ebook record available at https://lccn.loc.gov/2024041625

DOI: 10.1093/9780190064419.001.0001

Hardback printed by Sheridan Books, Inc., United States of America

To Jeff and Peter. We're sorry and thank you.

Contents

Introduction

> *Not ideas, but material and ideal interests, directly govern men's con-*
> *duct. Yet very frequently the "world images" that have been created by*
> *ideas have, like switchmen, determined the tracks along which action*
> *has been pushed by the dynamic of interest.*
>
> —Max Weber[1]

Legend has it that in 1945, in a speech commemorating victory over Japan, Army Air Corps General Hap Arnold declared, "We have just won a war with a lot of heroes flying around in planes. The next war may be fought with airplanes with no men in them at all. . . . Take everything you've learned about aviation in war and throw it out of the window and let's go to work on tomorrow's aviation. It will be different from anything the world has ever seen."[2] Arnold went on to lead the US military's newest service, the Air Force, as the intellectual father to a fraternity of airmen who loved airpower more than anything else in the world. This was a service of leather-jacketed pilots made of the right stuff, united by a desire to slip "the surly bonds of earth, and dance the skies on laughter-silvered wings."[3] How strange, then, that a service so enamored with and entwined with the love of flight would also see the greatest proliferation of unmanned systems over the next sixty years. How prescient, also, that a man who spent his life fighting and leading large wars of attrition would foreshadow the rise of unmanned technology in the future of American conflict.

Unmanned systems—a grab-bag grouping of automated, remotely operated, and autonomous platforms, robots, decoys, sensors, missiles, torpedoes mines, bombs, and satellites—caught the imagination of America and its warfighters long before Hap Arnold. From early mines in the Revolutionary War to Civil War torpedoes, the desire for technology to augment or even substitute for humans has been a luring odyssey for the American military. Viewed from this vantage point, the rise of American unmanned systems seems predetermined, an almost banal tale of inevitable technological progress. But the reality of the rise of unmanned technology within the American military is not a science fiction fantasy, nor was its trajectory inevitable. Instead, the rise of unmanned technology in the US military is a

The Hand Behind Unmanned. Jacquelyn Schneider and Julia Macdonald, Oxford University Press.
© Oxford University Press (2025). DOI: 10.1093/9780190064419.003.0001

complex saga of beliefs, identities, and policy entrepreneurs, all hidden in a handful of budget lines, between tanks and fighter jets and aircraft carriers—defense budget annotations that represent decades of critical junctures, path dependencies, and exogenous shocks.

This book isn't simply a glossary of today's unmanned systems or a recounting of what unmanned systems the United States bought or used. Instead, this book explains *why* the United States invests in these technologies in order to explain what systems make it into budgets and, eventually, onto battlefields. Despite the rhetoric of inevitability that surrounds unmanned systems, our book shows that the rise of unmanned systems in warfare was never inevitable, and in fact, there are still significant uncertainties about the trajectory these systems will (or should) take. Why, for example, is the contemporary arsenal dominated by aerial unmanned systems? Why did the United States in the past 20 years invest so heavily in remotely controlled platforms versus the munitions that dominated earlier developments? Why was the Air Force the leading investor in unmanned platforms on the battlefield—despite the influence of manned pilot culture? And, perhaps most puzzling for today, why was the US unmanned arsenal, dominated by remotely controlled unmanned aerial platforms, so ill-prepared to support a war in Ukraine that featured an army of small drones, loitering munitions, and ground launched missiles?

Previous examinations of unmanned technology have tended to focus on an artificially binary choice between manned and unmanned platforms. However, the truth is that unmanned technologies are diverse—used for different types of missions, within different domains or services, and with varying levels of autonomy. It is the variance in unmanned technology that reveals the most interesting trade-offs in unmanned trajectories. It is not *why not* drones but instead *why* platforms like remotely piloted aircraft instead of myriad other unmanned paths that lead to satellites, mines, torpedoes, cruise missiles, ballistic missiles, decoys, underwater unmanned vehicles, underwater ground vehicles, unmanned surface ships, or unmanned vehicles. The paths toward autonomy not taken, abandoned, or even stumbled into can have far more explanatory power about the shape of unmanned arsenals than simple manned-unmanned trade-offs.

The argument put forward in the following pages therefore questions strictly rational explanations for the development of unmanned technologies and traces the role of beliefs and identities within the US defense community to help explain the proliferation of these systems.[4] The dominance of unmanned aerial vehicles today was hardly preordained but instead represents a series of choices made by human individuals and organizations, which we argue are often motivated by their beliefs and identities rather than a purely rational equation about military effectiveness. This book, therefore, is just as much about the invisible human hand as it is about unmanned technology. Making that human hand more visible in the story about unmanned technology reveals something about how states choose the ways in which they fight wars. It helps us to better understand the relationship between human nature, organizations, states, and the conduct of war.

What Do We Mean by the Rise of Unmanned Technology?

While the dominant puzzle of this book is why platforms like the Predator and the Reaper dominated US contemporary unmanned arsenals, it should be clear that this book is not just about drones. Instead, we use the term "unmanned systems" to include a wide array of technologies that remove the human from the engagement, often by employing automation or autonomy (sometimes crudely). They are therefore systems that are uncrewed, uninhabited, and not necessarily controlled. These systems may navigate independently, operate without human intervention, or update tasks or targets without human control. They are not necessarily "smart" or autonomous, and in fact historically most are remote-controlled or "dumb" systems that remove the human from operations but without substituting machine decision-making for that independence. Unlike studies of autonomous weapons that define the genre by the ability to operate independently, we view autonomy as a variable characteristic of these systems, not necessarily a requirement. This means we include crude systems that operate without human control and with limited or no ability to change their operation, as well as what Paul Scharre describes as machines that "sense the environment and act" in an "immediate and linear" manner, more complex automated systems that "consider a range of inputs and weigh several variables before taking an action," and finally "goal-oriented" systems that are fully or semiautonomous.[5]

Our conception includes both platforms and munitions and extends across all warfighting domains. We look at unmanned systems that perform support, defend, and even strike missions. This expansive definition means that unmanned systems include some automated systems that are extremely rudimentary. For example, unmanned listening posts, mines, and torpedoes have been in military arsenals for hundreds of years, while satellites, missiles,[6] and unmanned aerial vehicles didn't make a significant advent until the digital age. More recently, with the integration of machine learning, unmanned arsenals have expanded to include unmanned underwater and surface vehicles, unmanned ground vehicles, and robotics. The wide array of technologies that fall under this conception is another reason that our study on the rise of unmanned technology is uniquely interesting. We focus on the variance within this group and, in doing so, identify a series of different potential trajectories that the United States took or could have taken in its investment in unmanned systems.

We identify five characteristics that define the trajectories of US military investments in unmanned systems and, ultimately, determine the winners and losers in the rise of unmanned technology within the Department of Defense. First, the choice to either invest in unmanned platforms (uninhabited airplanes, ships, or tanks) or munitions (torpedoes, missiles, bombs, or mines) represents a significant bifurcation in American unmanned trajectories. Historically, unmanned munitions were easier to engineer and less costly than unmanned platforms, but they were also less optimal for missions that required persistence, last-minute updates, or situations with high risk for either fratricide or collateral damage. Especially before the advent of radio control or microprocessors, these munitions were also offensively inclined and ceded autonomy to the weapon for strike, often doing so with little machine intelligence to differentiate friendlies from enemies or the probability of effectiveness.

This leads to the second fork in the road for unmanned investment—decisions about levels of autonomy. The Reaper and Predator, for example, are remotely controlled, while modern missiles—with many different types of initial and terminal guidance—are often semi- or even fully autonomous. Unmanned reconnaissance, whether it was listening posts, satellites, or unmanned aerial vehicles, was initially automated, but the advent of long-range digital control made these systems more capable of dynamic human retasking. These examples highlight that the first few centuries of unmanned systems decisions about levels of autonomy were often driven by technical capabilities. However, in the postmicroprocessor age, decisions about autonomy (especially for armed systems) are driven by a far more complex set of beliefs and perceptions that include discussions about mission effectiveness, risk, ethics, and often the law. As technology became more capable, paradoxically, human agency over the use of autonomy also became more discriminatory.

This has knock-down effects for yet another trajectory split—the decision to use unmanned systems for different missions and at different levels of warfare. Intelligence, for example, is often fully autonomous or at least automated, while strike may be more likely to be remotely controlled. Some missions, for example, those considered dull, dangerous, or dirty, may lead to investments in unmanned systems that are expendable, while other missions that prioritize control and precision may lead to more expensive platforms and munitions with complicated and high-cost logistical and communications support. Related, decisions about levels of autonomy and missions are often driven by how decision-makers prioritize different levels of warfare, most notably strategic or tactical. Historically, many US unmanned systems were designed for and controlled at the strategic level despite significant experimentation within combat. Perhaps in one of the most dramatic trajectory splits since the end of the Cold War, contemporary unmanned proliferation occurred at the tactical and operational levels of warfare despite a previous half-century focus on strategic unmanned systems.

Finally, unmanned systems may vary based on the domain in which they operate—air, land, sea, or space—as well as the services that invest in and operate these systems. While some of these domains are easier to engineer for unmanned systems (air, for example, has less clutter than the ground), others present opportunities for new missions that may be worth the engineering challenge if deemed an essential mission for the owning military service. All of these decisions about platforms versus munitions, operating domains, levels of autonomy, or mission purpose reflect assumptions about principles of warfare. How the US military privileges different tenets of warfare (e.g. speed, agility, persistence, precision, firepower, or cost) leads to different decisions for unmanned systems. These tenets of warfare also represent a constantly evolving relationship between technological possibility, organizational identity, and strategic beliefs.

The Inevitability of an Unmanned Future

We are hardly the first scholars to focus on unmanned technologies. Scholars and practitioners have been drawn to unmanned systems, curious (if not evangelistic) about how these technologies will transform how militaries fight. Underlying this curiosity is an assumption that investment in unmanned technology will improve the military effectiveness of those who employ these new technologies. Like military revolutions that came before—from the longbow to the development of gunpowder, steam, the telegraph, the railroad, and mechanization—states that adapt quickly to an unmanned future will be vastly superior to those that fail to do so.

Based on this assumption that unmanned technologies create military effectiveness, the key determinant in the diffusion of these technologies becomes not variation in state (or organizations') interest to invest and acquire these technologies but rather the capacity of states to build these systems. As Andrea and Mauro Gilli argue, while "the spread of drone warfare could redistribute military power at the global level and eventually promote instability and conflict . . . IR scholars have largely underestimated both the technological challenges of developing advanced weapon systems . . . and the infrastructural support they require."[7]

Despite how prevalent the assumption is about unmanned inevitability and capability within the literature on unmanned systems, these same advocates struggle to articulate how the characteristics of unmanned technology lead to revolutionary changes in conflict.[8] In a 2015 interview at the Center for Strategic and International Studies, the deputy chief of naval operations for warfare systems was asked about the Navy's support for the Unmanned Combat Air System. The admiral replied: "I'm a firm believer that we need that unmanned aircraft in our fleet and it will make the air wing that much better,"[9] but he went on to say that he wasn't sure how or why the unmanned system would make the Navy more effective. Similarly, when we conducted a series of interviews of ground personnel (joint terminal attack controllers) about their preference for manned or unmanned aircraft, we found that despite a huge preference for manned aircraft, there was a general belief in the inevitability of unmanned strike aircraft.[10] As one Marine noted in frustration to a room of controllers concerned about the proliferation of unmanned strike capabilities, "Guys, it doesn't matter if we think manned or unmanned is better. Unmanned is the future and we better get on board."[11]

However, as we discuss in Chapter 3, an examination of manned and unmanned system characteristics highlights advantages and disadvantages of both. Many of the characteristics proponents cite in favor of unmanned systems in future warfare, such as precision, loiter, cost, maneuver, and force protection, either are not unique to unmanned systems, come with important trade-offs, or vary based on the unmanned mission. Optimizing force protection, for example, may decrease risk to servicemembers but introduce high logistical costs if controlled remotely. Strictly capacity-based arguments therefore struggle to explain the uneven pattern of investment in unmanned platforms across the US armed services and cannot explain why some unmanned characteristics are invested in over others or how these characteristics create unique or revolutionary advances on the battlefield.

Our Argument in Brief

If the inevitability of unmanned systems is an assumption, then what is driving this belief in an unmanned future? To answer this question, we build on the work of previous scholars who emphasize the importance of constructed beliefs in explaining the ultimate use and proliferation of technology. Nina Tannenwald's research, for example, shows that materialist explanations for the nonuse of nuclear weapons are insufficient. Instead, it is beliefs, and in particular a "nuclear taboo," in the minds of decision-makers that best explain American nuclear restraint.[12] Similarly, Matthew Ford's examination of small arms in the twentieth century also highlights the role of perceptions in weapons development. He argues it is impossible to understand small arms innovation in the West without taking into account the way in which "different representations of a technology are constructed by different social groups involved in the design, production, and use of an artefact."[13] Finally, Rebecca Slayton shows how the constructed arguments made by two different professional communities—physicists and computer scientists—shaped investments in ballistic missile defenses during the Cold War.[14]

In the following chapters, we argue that how people and organizations think about unmanned technologies—that is, the beliefs that they hold about them—is crucial to understanding the future use and proliferation of these new military systems. By beliefs, we mean quite simply "that which we hold to be true" about the world around us. According to Johnathan Renshon, beliefs may be "propositions about causal relationships or fundamental assumptions about the way the world operates."[15] However broadly or narrowly conceived, beliefs are crucial in providing a guide to human behavior and to explaining the choices that individuals and groups make.

We argue that, first and foremost, service and occupational identities frame new technologies, shaping which unmanned systems services invest in and which are largely ignored. Unmanned systems that support armed service (the Army, Navy, Air Force, and Marine Corps) identities are more likely to be adopted, while unmanned systems that threaten core identities are more likely to be shunned or ignored. Further, these identities are extremely sticky, which means they represent the status quo trajectory for unmanned systems given no exogenous shocks, interservice competition for unmanned resources, or competing beliefs.

These core status quo identities interact with two key beliefs to help explain the trajectory of unmanned proliferation within the United States. The first belief, military revolutions and technological determinism, is the belief that technology exists within a linear understanding of history in which technology punctuates equilibrium to create revolutionary advances in military effectiveness. Unmanned systems are a part of that linear progression, as a component of the most recent information technology revolution.[16] According to these beliefs, the primary agent of change is technology itself. It is therefore the responsibility of the United States to harness the power of unmanned technologies to leapfrog adversaries by creating campaigns of speed, situational awareness, and decisive advantage. The second set of beliefs is about casualty aversion and force protection. It holds both that the US public is casualty intolerant and that their opinion is important for the military to achieve strategic success.[17] Public opinion about the loss of troops constrains decision-makers and influences the choice of military tools on the battlefield. By removing US personnel from the battlefield, unmanned technologies provide a technological solution to the constraints decision-makers believe are imposed by the American public's casualty intolerance.[18]

How do these beliefs and identities impact the proliferation of unmanned technologies in the United States? These beliefs become more or less powerful depending on (1) the salience of the beliefs, (2) the presence of institutional or individual champions for unmanned technologies, and (3) the emergence of critical junctures that permit new ideas to enter the policy space. These three conditions need not always be present for beliefs about unmanned proliferation to matter, but they increase the likelihood that these beliefs will translate into meaningful policy change, especially when and if unmanned systems challenge core service and occupational identities.

First, building on previous scholarship on norm proliferation, we argue that these beliefs become more consequential to unmanned proliferation when they graft onto existing understandings about war, international relations, and technology.[19] For example, Richard Price argues that the international effort to ban land mines was successful due to proponents' ability to borrow or graft from taboos on other weapons—in this case, chemical weapons.[20] This pre-existing normative framework made a ban on land mines far more acceptable an outcome than would have been the case in the absence of other weapons prohibitions. Thus, to the extent that beliefs about unmanned technology resonate with common understandings of how the world works and provide a familiar solution to an emerging problem, they can provide a roadmap for use that paves the way for further investment in these platforms. Beliefs are most salient and most likely to affect unmanned outcomes when they are lessons learned from combat. The visceral experience of warfare and the lessons that young military members take from these early experiences in warfare create strong beliefs that are more likely to affect outcomes than those beliefs that are created in the staffing bowels of the Pentagon.

Second, these beliefs become more influential to policy when they are championed by either an institutional body or an enterprising individual. Pre-existing institutions will respond to technologies based upon whether, and in what ways, that technology will benefit the material interests of the organization. Armed services, for example, integrate beliefs into doctrine and strategy documents that then marshal and prioritize resources for certain technologies. When beliefs about unmanned systems and their purpose are codified by these documents, they are more likely to lead to defense investments and procurements. In contrast, the defense industrial base works from the outside, advocating for some unmanned technologies that benefit their profit margins while using lobbying organizations to block or change other technologies that are less likely to benefit their companies. In doing so, defense organizations draw on and magnify beliefs about casualty aversion and the inevitability of the unmanned revolution and play on service identities in ways that can help them advocate for their agendas within the policy-making process.[21]

Individuals are crucial means by which beliefs become translated into policy. Existing work on military innovation already highlights the important role of senior military officers or civilian leaders in pushing military organizations to change.[22] More generally, there is a large and growing body of literature on policy entrepreneurs that stresses the importance of individual agents in creating openings for and enacting policy change. This is especially likely to occur when individuals have the power, resources, and opportunities to push their beliefs forward, build networks, and generate support for their policy agenda. This group could include senior decision-makers who have the executive authority to translate their beliefs about unmanned weaponry into action and nongovernmental analysts and academics who have established credibility and garnered the trust of those in positions of power.[23]

Finally, these institutional and individual beliefs are more likely to have an impact in the policy community when there are windows of opportunity or new ideas to gain traction. These windows can open at predictable intervals with budget renewals as well as during regularly scheduled election cycles. At other times internal domestic crises or external shocks caused by shifting geopolitical dynamics and adversary threats can puncture the political equilibrium and generate a need for innovative solutions and ideas to address new or pre-existing problems. These junctures create openings for institutions and individuals to push forward their beliefs and policy agendas in an environment that is amenable to change. Once adopted, these beliefs can become embedded in policy and create new path dependencies that shape the trajectory of weapons development.[24]

Taken together, we argue that the salience of beliefs, the promotion and ultimate internalization of these beliefs by enterprising individuals and institutions, and the opening of windows of opportunity create the means by which beliefs become embedded in decisions about unmanned weapons proliferation.

Methodology

The focus of this book—unmanned proliferation—includes not only the acquisition and use of unmanned systems but also their research and development. We explore this proliferation by detailing acquisition programs, research and development initiatives, and unmanned operations. Sometimes we can compare unmanned proliferation by a total amount of budget resources—for example, programs with $10 billion versus smaller programs in the millions. However, we go beyond budgets to look at the overall success of these programs. Were they able to create operationally capable systems? Were they used in training? In combat? How effective were they in campaigns? How extensively were they adopted across the military services?

In this book, we argue that unmanned proliferation is as much about the beliefs that individuals have about unmanned technologies as it is about the capacity of states to create these new systems. The capacity explanation is what we view as the dominant argument in current practitioner and scholarly discussion of the rise of unmanned technology—that US investment in unmanned weaponry is a strictly rational cost-benefit calculation of military effectiveness and that the trajectory of development is based solely on the ability of the state to create institutions and technological capacity to create the systems. We explore this capacity argument in detail in Chapter 3.

We spend the rest of the chapters expanding on alternative ideational explanations for the rise of unmanned technology. While these identities and beliefs do not operate strictly independently of one another, we explore their role in the proliferation of unmanned weapons separately: first, the belief in military revolutions; next, the casualty aversion belief; and finally, identities through which core constituencies view these beliefs and, subsequently, emerging technologies. These beliefs are held by a variety of actors within the US defense ecosystem—defense technocratic elites, academics, congresspeople, presidents, the defense industrial base, and the armed services itself. These different actors hold positions of prominence in the decision-making about unmanned systems, but they do so unequally and at disparate moments. Part of our exploration is identifying when and how each of these actors mattered in the rise of unmanned technology in the US Department of Defense.

We access the beliefs of these actors primarily through historical process tracing of doctrine, memos, and public writings, as well as interviews. We trace the origins of beliefs in casualty aversion to the Vietnam War and beliefs in military revolutions back to the Soviet Union during the Cold War. Through elite interviews and historical case studies, we show how these beliefs—though different in origin and operation—became internalized and promoted by key individuals. Finally, to show how these beliefs have come to influence the trajectory of unmanned proliferation, we show how crucial critical junctures created by exogenous shocks to the US system—Vietnam, the end of the Cold War, 9/11—created windows of opportunity for these beliefs to resonate with policy makers and to provide familiar solutions to oftentimes new and unforeseen problems. In providing these solutions, we document the long-term path dependencies that explain old and new patterns of unmanned investments.

How the Story Unfolds

The remainder of the book is devoted to telling a story about the rise of unmanned technology in the contemporary US Department of Defense. However, to understand where our story begins, the Chapter 1 provides background on how we got here: the evolution of unmanned systems from the early balloons of the nineteenth century through the Cold War and to investments in unmanned technologies today. This chapter serves as both a primer to inform the argument that develops in the remainder of the book and a clear detailing of the outcome of interest—namely, the proliferation of unmanned systems in the US military inventory.

The third chapter develops the theoretical framework for the book, exploring the rational choice explanation for the proliferation of unmanned technology and explaining why that argument leaves important questions on the table. We then introduce the role of identity and beliefs into the story to help provide context for structural or capacity approaches. We detail existing scholarship that illustrates powerful belief-based arguments and use this as a springboard to introduce the key beliefs that we believe help explain current patterns of unmanned proliferation in the US military—military revolutions and force protection (or casualty aversion)—and discuss how they inter-act with service identities. Finally, we set out hypotheses about how these beliefs—whether they emerge organically from experience or are cultivated and strategically promoted by the technocratic elite—are expected to influ-ence the trajectory of unmanned weapons development. This discussion sets the scene for the remainder of the book.

The fourth chapter focuses on the development and propagation of beliefs in military revolutions and technological determinism. Central to this belief's proliferation, and one of the major characters in our story, is the director of the Office of Net Assessment, Andy Marshall. Andy Marshall is the ideal policy entrepreneur and the chapter traces the beliefs he pro-liferated through the networks of influence he built. We detail the belief's emergence at the end of the Cold War, its cascade through the services in the 1990s, and the final internalization (with hiccups) after 9/11. In doing so, we identify Marshall's proteges who advocated for unmanned systems and advanced the narrative of military revolutions, all in the hope of creating networks of unmanned systems that increased situational awareness and decreased the time between sensor and shooter. Despite the strength of the military revolution belief over time and the innovative means of influence exercised by Marshall, we also identify limitations that hindered the belief from making a real difference to the ultimate trajectory of unmanned systems.

This leads to our fifth chapter, which traces the belief in public casualty aversion from experiences in Vietnam to force protection doctrines in the 1990s and early 2000s and finally to campaigns in Iraq and Afghanistan. These beliefs in the need to assuage public casualty aversion lead to investments in unmanned systems that privilege risk mitigation and, ultimately, precision and persistence. Unlike beliefs in military revolutions, casualty aversion beliefs are propagated unconsciously, passed down in combat experience from Vietnam, through the Powell and Weinberger doctrine, and finally making their way into the theories of victory for what have become long wars of political attrition in places like Afghanistan. We explore how these ideas affected doctrine, operations, and their subsequent acquisitions programs to explain how these beliefs shaped many of the remotely controlled unmanned aerial vehicles that dominate today's trajectories.

Finally, our sixth chapter looks at service and occupational identities and examines how the desire to preserve these identities influenced the trajectory of unmanned systems in each of the armed services. We explore in depth how branch identities and occupation identities, like pilots, interact with these service identities to create sometimes paradoxical investments in unmanned systems. Most importantly, in this final empirical chapter we examine how service and occupational identities mediate beliefs, ultimately impacting when and how these beliefs impact unmanned trajectories.

The final chapter synthesizes the three empirical chapters to weave a narrative about the push-and-pull forces between beliefs and identity that culminate in decisions to invest in unmanned technologies. In this concluding section, we also delve into the contemporary puzzle of why the United States' unmanned arsenal was so ill-equipped to support the Ukrainian conflict and how the use of unmanned systems in contemporary conflicts will influence the future of the US unmanned arsenal.

Notes

1. Max Weber, "Social Psychology of the World's Religions," in *Essays in Sociology* (New York: Oxford University Press, 1946), 280.
2. As quoted in Thomas McCurley, "Molding Generations of Air Force Superiority," Afcent.af.mil, March 20, 2011, https://www.afcent.af.mil/News/Commentaries/Display/Article/223247/molding-generations-of-air-force-superiority/.
3. John Gillespie Magee, "High Flight," https://nationalpoetryday.co.uk/poem/high-flight/.
4. For more on this line of analysis, see Judith Goldstein and Robert Owen Keohane, eds., *Ideas and Foreign Policy: Beliefs, Institutions, and Political Change* (Ithaca, NY: Cornell University Press, 1993).
5. Paul Scharre, *Army of None: Autonomous Weapons and the Future of War* (New York: W. W. Norton & Company, 2018), 28–30.
6. We differentiate in this exploration between ammunition or munitions with no autonomy that operate solely by kinematics (i.e., throwing a rocket or a dumb rocket or bomb with no guidance or targeting) and those that guide or detonate via automation or autonomy. See e.g. distinctions between guided and unguided missiles in Jacob Neufield, *The Development of Ballistic Missiles in the United States Air Force, 1945–1960* (Washington, DC: Office of Air Force History, United States Air Force, 1990).
7. Andrea Gilli and Mauro Gilli, "The Diffusion of Drone Warfare? Industrial, Organizational, and Infrastructural Constraints," *Security Studies* 25 (2016): 51–52; Matthew Fuhrmann and Michael Horowitz, "Droning On: Explaining the Proliferation of Unmanned Aerial Vehicles," *International Organization* 71, no. 2 (2017): 397–418.
8. Antonio Calcara et al., "Why Drones Have Not Revolutionized War: The Enduring Hider-Finder Competition in Air Warfare," *International Security* 46, no. 4 (2022): 130–171.
9. "The Future of Naval Capabilities," Center for Strategic and International Studies (CSIS), filmed August 5, 2015, at CSIS, Washington, DC, https://www.csis.org/events/future-naval-capabilities.
10. Julia Macdonald and Jacquelyn Schneider, "Battlefield Responses to New Technologies: Views from the Ground on Unmanned Aircraft," *Security Studies* 28, no. 2 (February 2019): 216–249.
11. Interview at Naval Base Coronado, anonymous. As cited in Macdonald and Schneider, "Battlefield Responses to New Technologies."
12. Nina Tannenwald, *The Nuclear Taboo* (Cambridge: Cambridge University Press, 2007).
13. Matthew Ford, *Weapons of Choice* (Cambridge: Cambridge University Press, 2017), 8.
14. Rebecca Slayton, *Arguments That Count: Physics, Computing, and Missile Defense* (Cambridge, MA: MIT Press, 2012). On the importance of beliefs in explaining the proliferation and use of military technology, see also Donald Mackenzie, *Inventing Accuracy* (Cambridge: Cambridge University Press, 1993); Adam M. Jungdahl and Julia M. Macdonald, "Innovation Inhibitors in War: Overcoming Obstacles in the Pursuit of Military Effectiveness," *Journal of Strategic Studies* 38, no. 4 (2015): 467–499.
15. Jonathan Renshon, "Stability and Change in Belief Systems: The Operational Code of George W. Bush," *Journal of Conflict Resolution* 52, no. 6 (2008), 822. See also Goldstein and Keohane, eds., *Ideas and Foreign Policy*, 8–11.
16. Clifford J. Rogers, ed., *The Military Revolution Debate: Readings on the Military Transformation of Early Modern Europe* (Boulder, CO: Westview Press, 1995); Steven Metz, *Strategy and the Revolution in Military Affairs: From Theory to Policy* (Darby, PA: DIANE Publishing, 1995); Geoffrey Parker, *The Military Revolution: Military Innovation and the Rise of the West, 1500–1800* (Cambridge: Cambridge University Press, 1996); David Parrott, *The Business of War: Military Enterprise and Military Revolution in Early Modern Europe* (Cambridge: Cambridge University Press, 2012); Colin S. Gray, *Strategy for Chaos: Revolutions in Military Affairs and the Evidence of History* (London: Frank Cass, 2002); Steven Metz and James Kievit, *Strategy and the Revolution in Military Affairs: From Theory to Policy* (Carlisle, PA: Strategic Studies Institute, 1995); Wim Smit and John Grin, *Military Technological Innovation and Stability in a Changing World: Politically Assessing and Influencing Weapon Innovation and Military Research and Development* (Charlotte: Virginia University Press, 1992); Max Boot, *War Made New: Technology, Warfare, and the Course of History, 1500 to Today* (New York: Penguin, 2006); Allan R. Millett and Williamson Murray, eds., *Military Effectiveness*, vol. 2, *The Interwar Period* (Cambridge: Cambridge University Press, 2010); Michael O'Hanlon,

Technological Change and the Future of Warfare (Washington, DC: Brookings Institute, 2000); Jon Lindsay, *Information Technology and Military Power* (Ithaca, NY: Cornell University Press, 2020).

17. See e.g. John Mueller, *War, Presidents and Public Opinion* (New York: Wiley, 1973); Scott Gartner, "The Multiple Effects of Casualties on Public Support for War: An Experimental Approach," *American Political Science Review* 102, no. 1 (2008): 95–106; Jonathan D. Caverley, *Democratic Militarism: Voting, Wealth, and War*, Cambridge Studies in International Relations (Cambridge: Cambridge University Press, 2014).

18. See e.g. Carl Builder, *The Masks of War* (Baltimore, MD: Johns Hopkins University Press, 1989); Jacob Neufeld, George M. Watson Jr., and David Chenoweth, *Technology and the Air Force: A Retrospective Assessment* (Washington, DC: Air Force History and Museums Program, 2007).

19. Martha Finnemore and Kathryn Sikkink, "International Norm Dynamics and Political Change," *International Organization* 52, no. 4 (1998): 887–917; Stacie E. Goddard, "Brokering Change: Networks and Entrepreneurs in International Politics," *International Theory* 1, no. 2 (2009): 249–281; Amitav Acharya, "How Ideas Spread: Whose Norms Matter? Norm Localization and Institutional Change in Asian Regionalism," *International Organization* 58, no. 2 (2004): 239–275; Martha Finnemore and Duncan B. Hollis, "Constructing Norms for Global Cybersecurity," *American Journal of International Law* 110, no. 3 (2016): 425–479.

20. Richard Price, "Reversing the Gun Sights: Transnational Civil Society Targets Land Mines," *International Organization* 52, no. 3 (1998): 613–644.

21. See e.g. Eugene Gholz, Harvey Sapolsky, and Caitlin Talmadge, *US Defense Politics: The Origins of Security Policy* (New York: Routledge, 2008), 135; Gordon Adams and Cindy Williams, *Buying National Security: How America Plans and Pays for Its Global Role and Safety at Home* (New York: Routledge, 2010).

22. See e.g. Barry Posen, *The Sources of Military Doctrine: France, Britain, and Germany between the World Wars* (Ithaca, NY: Cornell University Press, 1986); Stephen P. Rosen, *Winning the Next War: Innovation and the Modern Military* (Ithaca, NY: Cornell University Press, 1994).

23. See e.g. John Kingdon, *Agendas, Alternatives, and Public Policies* (Boston: Little, Brown, 1984), 151. In his pioneering definition, Kingdon noted that policy entrepreneurs could be "in or out of government, in elected or appointed positions, in interest groups or research organizations" (122). See also Michael Mintrom, *Policy Entrepreneurs and School Choice* (Washington, DC: Georgetown University Press, 2000); Michael Mintrom and Philippa Norman, "Policy Entrepreneurship and Policy Change," *Policy Studies Journal* 37, no. 4 (2009): 649–667.

24. Kingdon, *Agendas, Alternatives, and Public Policies*; Mintrom, *Policy Entrepreneurs and School Choice*.

Chapter 1
The Rise of Unmanned Technologies

The future of war has a distinctive and revealing past
—Lawrence Freedman[1]

To explore the role of beliefs and identities in creating the contemporary US unmanned arsenal, we must first take a glimpse into the arsenal. The following chapter details the path the United States took toward today's inventory, tracing the development of unmanned systems through history from the early use of balloons in the nineteenth century to the Cold War and the campaigns in Iraq and Afghanistan. The chapter ends by turning to contemporary US investment in unmanned systems, looking at budget and procurement by warfighting domain, armed service, and mission. While this chapter strives to tell the story of how unmanned systems were invented, developed, and used from the early days of unmanned experimentation to current investments in unmanned systems, the question of why history unfolded in this way remains open and will be taken up in the following chapters.

First, what are the unmanned systems that our story follows? Most obviously, we are exploring military systems—platforms or munitions—that have some autonomy to accomplish their mission. That autonomy could exist in myriad ways, whether through autonomous navigation of an environment, the collection of information, decisions about employing weapons, or tracking targets. Unmanned systems span a range of autonomy, from those that are completely autonomous in both operations and employment, to other systems that are remotely operated by humans, to others that are fired by humans but have some semiautonomy to complete their mission.[2] They include platforms designed to return after missions as well as platforms and munitions that are designed for only one use (i.e., send only). They are platforms for reconnaissance—listening stations, satellites, and unmanned aerial vehicles—but also systems for logistics, communications, and strike. Some unmanned systems operate far from the initial manned hand-off—in different countries, across the globe, and even in space. But some unmanned

The Hand Behind Unmanned. Jacquelyn Schneider and Julia Macdonald, Oxford University Press.
© Oxford University Press (2025). DOI: 10.1093/9780190064419.003.0002

systems have much shorter tethers, within line of sight of a manned unit or platform. Some are large and expensive, while others are small and cheap. They operate across domains and in all the services. This diversity in characteristics is what makes the trajectory of unmanned systems so fascinating because at any point in history some of these characteristics could have dominated the others. Certainly, there are some characteristics that are tied to the physics of the technology itself—the difficulty, for example, in navigating through cluttered terrain or the challenge of controlling unmanned systems before the advent of wireless communications or microprocessors. But other characteristics of the unmanned inventory—size, service, mission— are tied very closely to decisions that humans have made in past budget cycles, military campaigns, and foreign policies.

Early Unmanned Systems to World War I

The story of the human quest for unmanned systems starts centuries ago. While today's discussion often reverts to tales about the Predator or Reaper and precision strikes in Afghanistan or Iraq, unmanned systems existed long before contemporary drones appeared on the modern battlefield. There are reports of balloons and mines in ancient China, while drawings from Leonardo Da Vinci in the early sixteenth century sketch out an unmanned "aerial screw," capable of vertical take-offs and landing without a pilot.[3]

While we generally associate today's unmanned systems with the air domain, unmanned systems really debuted as mines—on land and at sea. There are references to Chinese land mines from as early as the thirteenth century, crude vessels filled with an early form of gunpowder and ignited first with tricky fuses and later with rifle locks. These technologies diffused with gunpowder to Europe in the sixteenth century and were occasionally used to defend castles, buried in slopes along the castle walls and detonated on contact.[4] This technique and new fusing technologies were cemented by the French two centuries later with the invention of the fougasse and camouflet mines, which became standard practice in fortification defenses in France and later the United States.[5] The United States experimented with land mines during the Seminole Wars, hiding shells under the bodies of dead soldiers, but with disastrous results. In one particularly gruesome episode, an American land mine set off by a possum drew American soldiers out of their fort only to be attacked by a waiting group of Seminole Indians.[6]

It was not until after the Crimean War and the development of Nobel fuses, which featured a more reliable pressure-fusing mechanism, that land mines drew more interest from the United States. It was the Confederacy's quest for asymmetric tactics to defeat a much stronger Union military that really brought land mines to the United States. Future president of the Confederacy Jefferson Davis sent a delegation of military officers to observe the Crimean War. The officers were impressed by the Russians' use of mines in defensive fortifications and rejected moral qualms about mines, arguing instead that it was the attacker who made their necessity unethical—not the weapons themselves.[7] They blamed traditionalists in the American military for not understanding the utility of land mines, one writing that "professional men are rarely the ones to invent or the first to adopt" potentially revolutionary weapons.[8] It comes as no surprise, then, that when the Confederacy declared itself a sovereign nation and embarked on its "war against Northern aggression," some of these officers were quick to adopt land mines. Confederate General Rains (also the leader of the disastrous mine attempt against the Seminole Indians) mined both the battlefield and the surrounding town during the Battle of Yorktown, essentially setting the systems up as indiscriminate booby traps.[9] Though Rains justified his actions as a necessary means to defend "against an Army of Abolitionists, invading our country for the purpose avowed, of extermination," General Longstreet wasn't convinced about the indiscriminate use of land mines, especially in civilian areas. He sent Rains to the Navy where mines might be less likely to create indiscriminate violence against women and children.[10] As the war progressed and the outcome became more dire for the Confederacy, many of the initial ethical qualms about land mines softened and the Confederacy increasingly turned to the weapons to slow the Union advance. Their effectiveness created begrudging respect from the Union Army and even some adoption within the advancing ranks of Union General William Tecumseh Sherman. As Sherman wrote in June 1964, "I now decide that the use of the torpedo is justifiable in war in the advance of an army . . . but after the adversary has gained the country by fair warlike means, then the case changes entirely. The use of torpedoes in blowing up our cars and the road after they are in our possession is simply malicious."[11]

Naval mine warfare began a few centuries after its land mine antecedents with reports of Chinese naval mines and European "fireboats" in the sixteenth century. Naval mines were first used by Americans in the Revolutionary War.[12] Invented by David Bushnell, these early mines consisted of watertight kegs filled with gunpowder, fashioned with a charge designed to explode on contact, and then floated into rivers, guided at the whim of the current. These crude mines were first deployed during the war against the British frigate *Cerberus*. Although the mines were intercepted before reaching the frigate, they managed to kill three British sailors and sink the schooner that intercepted the mine. In an incident later immortalized in Francis Hopkinson's *The Battle of the Kegs*, the Americans tried again with their crude mines by floating the kegs into the Delaware River to break up a British blockade. The kegs so alarmed the British sailors that British forces shot blindly into the river, expending large amounts of ammunition. As Hopkinson's ballad recounts, "The cannons roar from shore to shore; The small arms make a rattle; Since wars began, I'm sure no man Ere saw so strange a battle. . . . Such feats did they perform that day Against those wicked kegs, sir, That years to come, if they get home, They'll make their boasts and brags, sir."[13]

Mines were an attractive unmanned option for a young America, an option that "reflected the public's desire to avoid a large navy."[14] Accordingly, naval mines were used more extensively over the next century—first in the War of 1812 (in largely unsuccessful efforts to sabotage British ships in American harbors)[15] and later during the Civil War, where Crimean War innovations with electrical charges allowed naval forces to remotely control explosions from shore.[16] However, these early unmanned systems languished as the United States built a professional Navy, increasingly concerned about the ethics of what was perceived as an asymmetric and uncivilized tool of naval warfare. Thanks to the innovations of stalwart proponents within the Confederacy (like General Rains), naval mines laid by the Confederate Navy, primarily placed in their own harbors and ports as a defensive tactic, sunk twenty-nine Union ships.[17] Even though these mines were powerful weapons against the Union Navy, they got little attention from the North. Famously, Union Navy Admiral Farragut referred to the crude unmanned munitions as "sneak" and "devilish" devices that he believed "unworthy of a chivalrous nation."[18]

The Civil War also saw the advent of a related naval technology, the torpedo. As opposed to the mine, the torpedo had its own propulsion, allowing the explosive to move toward targets without relying solely on currents. These early torpedoes began as remotely controlled unmanned systems, tethered to coastal defense sites by a cable extending up to two miles. The tethered control allowed these weapons significant manned intervention, making them uniquely adaptable and accurate compared to both mines and nonguided torpedoes. However, that same tether also limited the use of torpedoes in naval strategy. The tether was cumbersome and vulnerable to interference (both manmade and environmental). Further, these tethers were impossible to use from a moving vessel, significantly limiting their utility in larger naval strategy.[19] In perhaps the most famous escapade, Union naval officer Lieutenant William Cushing hand-deployed a torpedo from his boat, successfully sinking the Confederate ironclad *Albermale* but at great loss to Cushing's crew—twelve died in the engagement.[20]

Some of the limitations of these early tethered torpedoes were overcome by the development of a new naval platform, the submarine, able to launch a spar torpedo from underwater. Notably, the Confederate submarine *Hunley* logged the first ever submarine kill when it used a torpedo to sink the *Housatonic* sloop (the *Hunley* and its manned crew were also lost for reasons that remain unclear).[21] Thus, a guided munition, albeit crude, led to a whole new naval platform and novel underwater missions. It would not be the last time that unmanned munitions shaped the trajectory of manned platforms.

It was not until after the Civil War that developments in both propulsion and guidance made torpedoes reliable enough to produce and deploy in mass. The leading system of the day, the British Whitehead torpedo, had a maximum speed of approximately fifty miles per hour and a range of 5.6 miles—making the torpedo a formidable munitions rival to the large guns that dominated the period's naval forces. The first successful self-propelled torpedo attack against an enemy warship occurred in 1891 during the Chilean Civilian War, and by the end of the century, the US Navy had over two hundred torpedoes in its inventory, deployed on torpedo boats, cruisers, and even battleships—making the torpedo the first unmanned system to be normalized in combat.[22] Moving into World War I, torpedoes were a mainstay on many capital ships, and navies even experimented with smaller boats designed specifically to launch torpedoes.[23]

These unmanned systems were successful enough to influence Navy force design during the turn of the century and were important enough to threaten the prevailing views of naval warfare. As Katherine Epstein recounts in her exploration of torpedo development within the United States, "The result was a race for range between guns and torpedoes that raised the possibility that the entire system of tactics built around capital ships armed primarily with big guns would give way to one built around smaller vessels primarily armed with torpedoes."[24] Indeed, in 1903 the US Navy's chief of the bureau of ordnance wrote to the secretary of the Navy that "the most effective use of torpedoes is most likely to be obtained when they are used on vessels designed especially to use them, such as torpedo boats and submarine boats,"[25] Accordingly, the success of the new torpedo boats forced changes in naval strategy, including the development of weapons like counter-torpedo boat destroyers and new strategies to negate the danger of torpedoes to capital-ship navies.

Despite the impact of torpedoes on naval strategy, their proliferation was by no means inevitable, and arguments against the systems were vociferous, especially among those invested in big-gun battleships. Torpedoes enabled new, smaller-surface platforms and elevated the importance of the submarine—both ultimately budget competitors against the capital ships of the day. Perhaps it was how the torpedo undermined existing naval force structure that engendered such strong reactions from inside navies. As the torpedo became more effective, arguments against its use increasingly turned to morality. For example, the torpedo's advantages in range and earlier hand-off for control were sometimes used as an argument against the munitions for being "illegitimate" and "weapons of the weak."[26] These early discussions foreshadowed contemporary debates about the ethics of unmanned weaponry but also highlighted the complicated relationship between unmanned munitions proliferation and the subsequent effect on manned platform investments and overall force structure design.

While the naval domain may have led early experimentation and unmanned innovation, the air domain was not far behind. There are reports of Chinese balloons, used for lighting and signaling across the battlefield, from as early as the third century AD.[27] Many centuries later, accounts suggest Austria planned to use unmanned balloons loaded with explosives to augment short-range artillery in the 1849 siege on Venice.[28] Despite these early advances, forays to develop unmanned air systems largely stagnated until the advent of the airplane in the twentieth century. This was because balloons (like Bushnell's floating kegs that relied on river currents for guidance) were especially vulnerable to weather changes. These early systems, which could not be remotely controlled, were essentially fully autonomous and dumb—unable to guide to a target, respond to human requests for changes, or correct for environmental shifts. Militaries launching these unmanned armed balloons could not ensure the balloons wouldn't cause unintended harm to civilians or, with a change in weather, end up actually harming the friendly troops that launched them in the first place.[29]

This made unmanned balloons dangerous weapons for strike missions. But what about other missions? Despite the allure of overhead views of the battlefield, unmanned balloons in the nineteenth century were also useless for intelligence (though manned balloons were used for forward observation, communication, and mapmaking). Without automated cameras and recording devices, or a reliable way to retrieve untethered balloons, these balloons were of no use without a human passenger. With so many problems with control and limited utility as an intelligence asset, militaries abandoned their early unmanned balloon experiments.

Two inventions at the beginning of the twentieth century made unmanned systems more feasible. The advent of the aircraft and the development of wireless technologies for transmitting and receiving information facilitated better (and potentially untethered) control and introduced the ability (albeit limited) to send and receive information. These advances, coupled with the advent of World War I, provided an impetus for new experimentation with unmanned systems. From propaganda-dropping balloons to target-training aids,[30] aerial torpedoes,[31] and even an early exploration of unmanned systems for strategic bombing and surface warfare,[32] unmanned experimenters in World War I expanded the ways that unmanned systems could be used on the battlefield. Meanwhile, unmanned stalwarts like land and naval mines continued to dominate unmanned arsenals, despite questions about both the ethics and effectiveness of these weapons.

Mines were poised for a larger role in conflict by the time World War I erupted. In the decades after the American Civil War, the weapons played a large role in Russian conflicts in Turkey, China, and Japan. Naval mines, now equipped with TNT and outfitted with automatic anchors that fixed mine depth, wreaked havoc in ports and waterways. During the Russo-Japanese War, for example, over two dozen vessels were destroyed by rudimentary mines, demonstrating their utility not only for coast defense and deterrence but also as a cunning tool to pen enemies into ports.[33] Their proven combat utility made them a luring weapon for many European navies (despite significant debate about the morality and legality of their use in the 1909 Hague Convention). This was especially true for the Germans, who got a head start in mine development and submarine mine-laying techniques over an initially reluctant British Navy.[34] Successful mine operations like those in the Dardanelles Straits in 1915 convinced the British that mines were worthy weapons of war. The US focus was initially on coastal defense—hoping that the cheap technology would leave more room in the already tight Navy budget for the Navy at sea.[35] However, by the end of the war, a reluctant US Navy followed the German lead and developed a new naval mine with a better firing mechanism. At the end of the war, in concert with the British Navy, the US Navy placed over seventy thousand of these anchored mines in the North Sea.[36]

German torpedo innovations also influence US torpedo development. German submarine U-boats, armed with torpedoes, demonstrated the power of the weapons both to influence big-gun battleship engagements and to interdict commerce and logistics. Perhaps most notably, on September 22, 1914, a German U-boat sank three British armored cruisers, killing almost 1,500 British sailors.[37] Two years later, in the 1916 Jutland Peninsula engagement, the German torpedo threat potentially convinced the British naval contingent to turn away from the German fleet.[38] Over a three-year period, torpedo-armed German U-boats sunk millions of tons of arms, supplies, and troops in the Atlantic—notoriously sinking the ocean liner *Lusitania* and with it 1,200 passengers.[39] The American Navy took note and commissioned thousands of torpedoes; by the end of World War I the American Navy had its first ever torpedo factory.

While the Navy dominated unmanned development, World War I also featured the debut of credible unmanned air systems. In 1918 the US Army Air Service successfully tested the Kettering Bug, an unmanned aircraft laden with explosives that used a rudimentary mechanical system to fly predetermined routes. Though the system was never used in combat due to concerns about control and possible fratricide, it served as the genesis of missile development programs in the US Army and later the US Air Force.[40] The Navy also made significant forays into unmanned research and development during World War I. As early as 1918, the Navy demonstrated the ability to remotely control unmanned surface vehicles without a cable tether, allowing the Navy to control these systems from land, air, and underseas.[41] Meanwhile, in Germany, a new class of unmanned surface vehicles, laden with explosives and controlled via radio communication by an overhead aircraft, was reportedly used to ram British vessels near the Estonian coastline.[42]

Land mines, in contrast to naval mines or torpedoes, never found favor with the US military, despite the introduction of the tank during World War I and the subsequent use of antitank land mines by the German Army. The reticence to adopt land mines was in part a product of their historical record. In conflicts between the American Civil War and World War I, land mines performed dismally. American military onlookers delivered a pessimistic verdict on the technologies: land mines were ineffective and unjustified. As one Army report concluded, land mines were not only immoral but also "nearly as dangerous to friends as foes."[43] Leading into World War I, the US Army's engineer's field manual declared that "it is not permissible to plan such mines in any ground which is not obviously prepared for defense."[44] This stance changed little during World War I, and it wasn't until World War II that the US military developed its own land mine. However, unmanned land systems made some progress during this time period. Although they never made it into conflict, the early twentieth century featured new patents for tethered land torpedoes—miniaturized robot explosive-laden tanks designed to solve the impasse of trench warfare.[45]

Despite this increase in unmanned innovation and missions during World War I, most of these systems continued to play only a minor supporting role in conflict. This was for a few reasons. First, wireless remote control was still difficult. Only torpedoes and naval mines had enough maturation in control and guidance to play an operational role in warfare. Second, many of the systems in development were less effective than their manned counterparts, and less survivable. Faced with limited control, effectiveness, and ultimately the vulnerability of these systems to both enemy attack and mechanical failure, only naval mines and torpedoes made a significant showing during World War I.[46]

The real significance of these expanded missions during World War I lay in the experimentation that could occur, and failure that was acceptable, in what were otherwise nonessential missions. For example, the use of unmanned surface vehicles for targeting practice had only a modest impact on naval effectiveness. However, to use these systems remotely, the Navy had to learn how to automate the basic processes in a ship—from the steering to the boilers.[47] These innovations built a knowledge base for future unmanned development in combat missions.

The trajectory of growth and experimentation during World War I and the early interwar years had similarities to the earliest uses of unmanned systems in the nineteenth century. Like the early unmanned platforms, most World War I unmanned innovation occurred in the naval and air domains, and, like the previous centuries (and the future century), widespread use of unmanned systems for strike was limited by the ability to technologically control both the platforms and the precision of strike. Munitions—namely torpedoes and the rudimentary forefathers to cruise missiles—dominated unmanned platforms, though their proliferation was limited by the threat they posed to the preferred manned platform of the day, the gunship.[48] Finally, missions that would later become the bread and butter of unmanned systems (namely intelligence) were still largely impossible due to difficulties in transmitting information. The US military weighed these technologies for both their effectiveness and their morality, shunning unmanned systems like the land mine, which seemed to have limited utility and a questionable moral justification.

Unmanned Systems and World War II

Some of these technological limitations that stymied the proliferation of unmanned systems during World War I were solved by the next major conflict. World War II saw revolutionary advances in radar and wireless communications. However, even as these systems showed greater utility—especially with the advent of wireless control—many were still unable to demonstrate enough military effectiveness to overcome concerns with reliability, cost and logistical support.

The US Navy began seriously experimenting with unmanned systems after World War I, including with unmanned surface vehicles and torpedoes in wargames at Newport's Naval War College.[49] From the integration of unmanned surface vehicles and torpedoes in wargames at Newport's Naval War College[50] to the use of unmanned surface vehicles to clear transit areas of mines,[51] and even the commissioning of hundreds of torpedo-carrying drones,[52] the interwar Navy was serious about exploring the uses of unmanned systems. There are even reports of secret Navy development of unmanned fishing boats designed to be used as explosive-laden vessels and sent into Japanese ports (though there is no evidence that these were ever used extensively or successfully).[53]

Given this experimentation, it is interesting to note how little the Navy's leading unmanned systems—torpedoes or mines—advanced during World War II. Both of these munitions had demonstrated their utility in World War I, and US adversaries, including the Germans and Japanese, invested heavily in the systems during the interwar years. However, the US Navy's torpedo and mine programs after World War I stalled. Faced with declining defense budgets, the Navy moved the munitions to the bottom of their priority list, behind the new aircraft, submarines, and carriers the service was clamoring for. Essentially all torpedo development outside of Newport, Rhode Island, was shuttered. That meant that when World War II began, the Navy was left with "a tiny dribble of beautifully crafted torpedoes, barely less erratic than their World War I forefathers, produced by an organization corpulent, sluggish, and not so much consciously resistant to change as physically and emotionally unable to."[54]

Consequently, the arsenal of torpedoes with which the US Navy entered World War II, including the aerial torpedo Mark 13, the submarine-launched torpedo Mark 14, and the surface-launched Mark 15, were difficult to launch, struggled with guidance, and were outranged by the far superior Japanese torpedoes. Despite US Navy platform innovations to employ torpedoes, including both manned air platforms and submarines, the weapons were largely ineffective until after 1943.[55] Although torpedoes made little impact on major campaign outcomes, they did play an outsized role in asymmetric tactics. For example, on the surface, the Navy developed a fleet of manned patrol boats designed to launch torpedoes as a "cheap to build" option to defend islands like the Philippines.[56] While not a major factor in US naval successes, the patrol boat torpedoes made Japanese logistics more difficult and harassed submarines.[57] Later innovations in sonar as well as smart torpedoes like the Mark 24 and 28 also increased US capabilities underwater. These late-war torpedo innovations created the impetus for underwater, antisubmarine warfare that then escalated throughout the Cold War.

Like torpedoes, US naval mines had a delayed entry to World War II. Despite their tardy debut, by the end of World War II, the US Navy had invented new systems with smarter fusing mechanisms that could be deployed by both submarine and aircraft. These new naval mines became a pivotal part of the Pacific strategy with over five thousand mines laid near Japan, China, and the Southwest Pacific. They are credited with destroying at least sixty Japanese ships and paralyzing shipping into and out of Japanese-controlled territories.[58]

World War II also generated significant experimentation in the air domain, culminating in the integration of unmanned platforms into combat. In perhaps the most concerted use of unmanned aerial systems during the war, the Japanese launched 9,300 explosive-laden balloons—of which 361 eventually arrived in North America.[59] But the United States was not far behind the Japanese, and in fact the US Navy's employment of a squadron of unmanned aircraft in the Pacific to aid strike efforts heralded the normalization of unmanned units within the standard warfighting force structure. The United States also sent a handful of unmanned bombers against targets in Germany[60] and used drones so heavily for targeting practice that a British invention, the OQ-2 radioplane, led to the world's first mass-produced drone.[61]

Overall, these unmanned aerial vehicles were only marginally impactful to combat outcomes, no doubt because even the new developments in radar and electronic transmissions couldn't overcome the extraordinary difficulty of controlling these systems. For example, in Operation Aphrodite, B-17s were stripped down and equipped with twice the firepower of a normal B-17. Unfortunately, the "unmanned" B-17 could not take off safely on its own and so the plane required a pilot and copilot to take off from inside the cockpit, transfer control to the mothership, and then parachute from what was at that point a B-17 drone. It was a complicated and dangerous concept. In the end, Operation Aphrodite and the Navy version Anvil included only fourteen missions; four crewmembers were killed (including the older brother of future president John F. Kennedy) and no drones successfully bombed their targets.[62] The program was terminated, but General Spaatz ordered that a few of the platforms be left, displayed, and ostensibly ready to launch to "leave in the minds of the Germans the threat of robot attacks."[63]

Unmanned aircraft may have had little combat impact in World War II, but the war introduced a new bifurcation in unmanned aerial systems' trajectory. On one path, unmanned aerial vehicles designed to return to sender that required significant remote-control technology, and on the other path, munitions like rockets, missiles, and bombs that increased the handoff distance between human shooters, their weapons, and the target. The possibility of guided, semiautonomous aerial munitions had captured the imagination of leaders within the Army Air Corps during the interwar years. Building on the success of naval torpedoes, General Billy Mitchell argued for a similar aerial munition, opining that "the aerial torpedo, a radio-controlled airplane packed with explosives, would be an excellent weapon for an enemy to use."[64] Indeed, in the early 1930s the Army Air Corps attempted to adapt the World War I Kettering Bug into a radio-controlled aerial torpedo as part of a classified "Study No. 9" that concluded that despite the technological difficulties and potential expense, "unmanned radio controlled aircraft have important military uses as aerial torpedoes."[65] While aerial torpedoes never made it to the battlefield during World War II, Army Air Corps and Navy experimentation with guided bombs, using televisions and early radio control to try and improve the accuracy of otherwise dumb bombs, did lead to limited success.[66] The Azon (or azimuth-only) manually guided bomb used radio control to allow a pilot to guide the bomb after launch. It was unwieldy and only useful against a handful of targets, but over three thousand were dropped in Europe and the Pacific.

It was a German innovation, the V-1 flying bomb, that ultimately spurred US investment in long-range guided munitions. The V-1, a forefather for later developments in cruise missiles, featured a gyrocompass autopilot and a range of over a hundred nautical miles. Although the V-1 struggled with accuracy and was vulnerable to both low-cost barrage balloons and antiaircraft fire, the potential for guided bombs and rockets spurred US research into guided rockets.[67] The Army Air Force and Army Ground Force embarked on an intraservice rivalry to create and control this new class of weapons and subsequently developed a series of long-range bombs and missiles, using television, infrared, or radio guidance to improve accuracy. Ultimately, however, these weapons saw little use in World War II. The most sophisticated of these innovations, the JB-2, debuted too late,[68] and though over five thousand were produced, none were used. This, of course, wasn't simply a timing issue. Many senior leaders, especially in the Army Air Force, were unimpressed by the German V-1 and did not see the missiles as useful replacements for airplanes and pilots, which were battle proven and far more economical.[69] In contrast, the German V-2 bomb, a far longer-range precursor to modern ballistic missiles, captured the imagination of the Army Ground Force and went on to play a much larger role in subsequent missile investments after World War II.[70] In 1945 then Army Air Corps General Hap Arnold commissioned a group of California scientists to look at these new bombs and determine their feasibility and strategic impact. The group reported back that "due to improvements in aerodynamics, propulsion, and electronic control, unmanned devices will transport means of destruction to targets at distances up to several thousand miles."[71]

While most of the unmanned development occurred in the air and to some extent the naval domains, even unmanned land systems benefited from technological and operational innovation. The biggest adopters of unmanned land innovations were not Americans but instead the Soviet and Germany militaries; the American military, to some extent, replicated some of these technologies. The Soviet and German militaries entered World War II with significant improvements in land mines, with better fusing to counter the new and faster tanks that dominated the early years of the war. They deployed these extensively in Europe, significantly slowing down advancing armies.[72] The Soviet Army also experimented with unmanned land platforms, going so far as to develop a completely remote-controlled tank that they deployed in the Winter War in 1939.[73] Large-scale adoption of these unmanned tanks was limited by their ability to navigate difficult terrain and weather, but unmanned mine-clearing vehicles saw greater proliferation during World War II than any previous forays into land-based mine-clearing unmanned systems.[74] Many other countries also experimented with unmanned explosive ordnance systems, but in the end the systems were too difficult to engineer and employ and were never able to demonstrate a degree of effectiveness that outweighed their complexity and cost. For instance, one of the most technologically innovative unmanned explosive vehicles, the German Goliath, was "too big to be easily man-portable, yet too small to carry a large explosive charge, and was readily susceptible to small arms fire. . . . [A]lmost 5,000 . . . were produced, of which 3,797 were unused."[75] Further, unmanned systems like the Goliath required a heavy logistics tail, necessitating a team of five or six infantry men just to set the system into its launch position. For Americans appraising unmanned land systems at the end of World War II, robots could not create enough military effectiveness to justify their cost or their logistical tail. Meanwhile, perhaps the most cost effective and simple unmanned ground system—the land mine—was mired in debates about whether these mines were effective enough to overcome concerns about their morality.

World War II was a catalyst for unmanned systems—especially for munitions in the air and in the water. As manned platforms became more sophisticated and changed the way war was fought, there were also more opportunities for guided and smart(er) munitions. The advent of the German V-1 and V-2 introduced a new trajectory for unmanned aerial systems, with significant implications for the path of unmanned systems in the Cold War. Finally, up until this point in our story, the trajectory for unmanned systems has been primarily about the ability to create enough control to use these systems as strike platforms. However, the next big era in unmanned development will introduce a new mission—intelligence—that threatened the dominance of strike within US unmanned missions.

Unmanned Systems and the Cold War: From Experimental to Operational

During the Cold War, a series of technological innovations helped spur unmanned systems' transition from experimental to operational. Microprocessors, space-capable rocket engines, solid propulsion, and new camera and sensor technology created new missions for unmanned systems. Microprocessors miniaturized the large and unwieldy set of vacuum tubes required to control robotic systems and made it possible to move artillery and crude missiles from weapons of barrage fire and massive damage to long-range, precision-strike unmanned systems. The microprocessor also solved the final problem for remote-control guidance for aerial platforms, allowing these systems to play a role in Vietnam. Meanwhile, advances in missile propulsion increased the range, reliability, and launch options for rockets, creating an aperture for both strategic ballistic missiles and space. Finally, advances in sensors paired with microprocessors and rocket propulsion increased the opportunities of space and with it came a proliferation of space-based autonomous capabilities for communications, navigation, and intelligence.

However, technology cannot completely explain the growth of unmanned systems (or which unmanned systems proliferated or languished) during this period. Exogenous shocks, strategic environments, and bureaucratic interests also created increasingly entrenched but distinct path dependencies for unmanned systems across domains. There were important nontechnological forks in the road—for both investment and adoption. Most importantly for this period, in 1947 there was an important exogenous shock that fundamentally altered the trajectory of unmanned systems. The National Security Act of 1947, which reorganized the Department of War and US intelligence agencies and created the Air Force, catalyzed novel organizational competition incentives that, in turn, spurred an internal food fight for control of missile and unmanned technologies. This bureaucratic change interacted with technology to shift the course of unmanned systems in the US arsenal, creating competitions not only between unmanned and their manned alternatives but also among unmanned systems. In particular, this period reveals three important trajectories for unmanned proliferation: smart(er) munitions, strategic unmanned platforms, and finally tactical experimentation.

The Rise of Smart(er) Munitions

Perhaps the most remarkable growth for unmanned systems during the Cold War was in munitions: torpedoes, mines, ballistic missiles, and cruise missiles, all with better fusing logics, guidance, and propulsion. Munitions dominated unmanned systems investment after World War II. This was for a series of reasons, many technological. Munitions didn't need to return home after their use, were comparatively much simpler to design than unmanned platforms, and didn't have the same logistical tail required to launch, land, and maintain unmanned platforms. This made munitions easier to develop, comparatively cheaper than platforms, and far easier to operate. Further, advances in rocket propulsion technology translated across the missile inventory, making these systems more effective across the board, while technological developments for unmanned aerial or surface platform control were still largely confined to specific platforms and therefore saw only marginal improvements across the fleet. However, munitions that had an early lead—like mines and torpedoes—stagnated during the Cold War, while long-range nuclear-capable missiles and, to some extent, cruise missiles became the bedrock of US strategy to deter the Soviet Union.

Coming out of World War I, it was munitions within the naval domain that held the early lead. However, despite torpedoes and mines' centuries-long unmanned dominance, the systems fell out of favor during the Cold War. Their missions became more circumscribed and the technologies found themselves relegated to the bottom of the budget. The primary use of torpedoes by the Germans, attacks against merchant and logistical vessels, was not embraced by the US Navy, and the Japanese use of both aerial and underwater torpedoes for surface warfare didn't align with US aircraft carrier–focused strategies. This meant that US Navy torpedoes were focused on one mission—antisubmarine warfare—which wasn't an immediate priority for the post–Cold War Navy.

It was only the tit for tat of submarine warfare that emerged in the 1960s and the introduction of new Soviet nuclear and attack submarines that drove the Navy to invest in new torpedoes. By 1970, Secretary of Defense Marvin Laird—concerned about Soviet submarines on the coast of the United States—pressed the Navy to invest in a new torpedo, the Mark 48, that would provide "the speed, range, acquisition, or depth capability required for use against modern, fast, deep-diving submarines."[76] While the decision to invest in the Mark 48 was a political response to Soviet advances, the possibility of the new torpedo was a product of technological trajectories.

The advent of the microprocessor gave torpedoes more autonomy and less dependence on wire guidance. With better autonomous seeking and less need for a submarine-sonar hand-off, US attack submarines had more flexibility to maneuver during torpedo engagements as well as greater range and probability of kill.

Naval mines also saw little advancement during the Cold War, and up until the 1980s, the United States primarily relied on a variant of the World War II Mark 6 naval mine. This included a notable use of naval mines in Wonsan during the Korean War and, during the Vietnam War, to interdict Viet Cong travel on rivers and to close the port of Haiphong.[77] Later versions of mines introduced in the late 1970s were more advanced, and closer aligned with torpedoes. For example, the Mark 60 Captor mine used sonar to initially track the adversary and would then launch a Mark 46 torpedo to follow and ultimately attack enemy submarines. While never credited with a kill, the Mark 46s were ostensibly designed to block Soviet submarines from transiting the Greenland–Iceland–United Kingdom gap. It was followed by a more covert Mark 67 mine, designed to be launched by submarine.[78]

The biggest unmanned munitions during the Cold War were not those that dominated World War I or II but instead a technology that made a late but notable entrance: missiles. Coming out of World War II, there were forty-seven guided missile projects in development. However, a major slash in postwar spending meant that number needed to be winnowed, and interservice competition dominated the ultimate shape of that winnowing, creating path dependencies for guided munitions throughout the Cold War.[79] As Stine concludes in his examination of US missile development post–World War II, "The US Army believed that long-range ballistic missiles were logical extensions of artillery. The US Air Force contended that guided missiles, even ballistic missiles, were unmanned aircraft and should come under its jurisdiction. The US Navy pointed to the 16-inch guns on its battleships and to the naval aircraft on its carriers and combined the arguments of the other two services."[80]

The Air Force's guided munition trajectory got a jumpstart over its rivals as the new service quickly realized it needed to create for itself a defendable budget niche within the new missile genre. As early as the end of World War II, General Hap Arnold pushed for the JB-2 missile (an American version of the German V-1), eager to demonstrate the weapon in combat to "stake out another role for the AAF (Army Air Forces)."[81] A few years later as political efforts to create the Air Force were coming to a head, Major General Curtis

Lemay (who years later would stymie missile development that he worried threatened US bombers) effectively "peed" on what he saw as Air Force missile territory, asserting "the long-range future of the AAF lies in the field of guided missiles."[82] In its quest to wrestle control of missiles away from both the Army and the Navy, the Air Force needed to frame these technologies within the new service's identity. It did so by characterizing missiles both as an extension of airpower and as a strategic weapon—two characteristics foundational to the new service's identity. To win this narrative, the Air Force began a campaign to describe missiles as "pilotless" or "robot" "aircraft"— making a clear distinction from the unguided artillery of the Army (this also led to a brief moment in 1946 where the Army Air Force made a play for antiaircraft missiles, or surface-to-air missiles, by calling them "supersonic pilotless aircrafts").[83] This was a precarious balancing act for the Air Force, keeping the Army and Navy from co-opting missile missions while making sure the munitions wouldn't threaten the Air Force's core bomber constituencies.[84]

This Air Force internal and external competition for missiles manifested in two choices after World War II: the first, air-breathing cruise missiles, and the second, ballistic missiles. Initially these two choices were not clear competitors; instead, the cruise missile, which more closely resembled manned aircraft, seemed a logical first step in a technological progression that would, in turn, lead to the more fundamentally different ballistic missile. As one Airman recounted, "To people who had grown up with manned bombers before and during World War II . . . a cruise missile was a less painful and certainly a less abrupt departure from what they were familiar with than would be a totally alien ballistic missile. Those who favored the evolutionary approach to the creation of a new generation of weapons, predominantly missiles, were people to whom aircraft had a meaning as a way of life, a symbol, a preferred means of performing a military assignment."[85] Accordingly, in the first decade or so after World War II, the Air Force put its short-term eggs in the cruise missile basket.

That changed, however, not long after the debut of the Air Force. The dominant organization within the Air Force, Strategic Air Command, worried that cruise missiles (which looked similar to manned aircraft) threatened Air Force long-range manned bomber aspirations.[86] They opted instead to invest in the fundamentally different ballistic missiles but argued that the new ballistic missile guidance systems were too crude to provide the accuracy to completely replace the manned bomber. This was a bit of a chicken-and-egg argument as Air Force leadership, worried that missiles would threaten their manned bomber force, invested little in missile guidance systems until the advent of the hydrogen bomb made ballistic missiles far more strategically useful.[87] As late as 1953, the vice commander of the Strategic Air Command wrote that, "regardless of the missile program, it is the opinion of this headquarters that the continued advance in the art of manned flight to high altitudes and long ranges should be at all times a priority objective of the Air Force's development programs."[88] As Michael Brown recounts in his exploration of Air Force strategic bomber development, "Accurate bomb delivery gave bombers a competitive advantage over unmanned systems in American budgetary battles for many years."[89]

It took civilian intervention, first from President Eisenhower's famous Solarium Commission and then from senior civilians in Air Force research and development, to create a new protected entity within the Air Force, focused solely on developing ballistic missiles.[90] This top-down effort, led by enterprising Air Force general officer Schriever, galvanized ballistic development, and the budget for the weapons went from $3 million in 1954 to $2.15 billion in 1958.[91] Under Schriever's leadership, the Atlas missile program and its descendant, the Titan, sprinted forward technologically as congressional and executive pushes ensured a solid budget allocation (support no doubt spurred by the Soviet Sputnik moment).[92] In just a few short years, Schriever pushed Atlas and Titan past their Army competitors and cemented the Air Force's control over both intercontinental ballistic missiles (ICBMs) and the shorter-range intermediate-range ballistic missiles (IRBMs). This was good news for the Air Force's overall budget as fears that the Soviets were leading the United States in missile and space development ultimately drove Eisenhower's scientific advisory committee to recommend an increase in IRBMs from 60 to 240 systems and an increase in ICBMs from 80 to 600.[93]

Under the Kennedy and later the Johnson administrations, ICBMs continued to grow. Part of this was due to technological advancements; the development of solid propellants (an innovation by the Navy's Polaris program) made the Air Force's new Minuteman far more flexible for rapid response.[94] Ballistic missiles also had a strong advocate in Secretary of Defense McNamara, who continuously stymied Air Force bomber initiatives in favor of both Air Force ICBMs and Navy submarine-launched ballistic missiles.[95] However, this precipitous rise in Air Force ballistic missiles plateaued as arms control talks, first begun under the Nixon administration and then culminating in the Strategic Arms Limitation Talks (SALT) II under Ford, further capped significant quantitative (and some qualitative) development of ballistic missiles. Remarkably, the Minuteman II and III inventory of 1975 (550 Minuteman IIIs and 450 Minuteman IIs) remained the same for the next thirteen and a half years.[96]

The Navy also invested heavily in ballistic missiles during the Cold War. Initially, this was part of a joint effort with the Army. However, an early test of liquid propellant IRBMs on a mock aircraft carrier led to catastrophic damage to the carrier and ultimately soured the Navy on joint ballistic missile development. Instead, the Navy went its own way in ballistic missile development, devoting its efforts to a solid propellant ballistic missile that could be deployed not on an aircraft carrier but instead on the new nuclear submarines being pushed by a bulldogged Admiral Rickover.[97] The success of the Navy's submarine-launched ballistic missile Polaris program cemented Rickover's vision of a strategic nuclear submarine, ensuring the Navy's role in strategic deterrence and solidifying the power of the submarine community within the larger Navy.[98] Like the Air Force's ballistic missiles, the Navy's submarine-launched ballistic missiles also plateaued in the 1970s as arms control treaties began to limit the deployment and development of these nuclear systems.

This led to an aperture for cruise missiles, which skirted the bounds of the arms control treaties. Early development of cruise missiles had initially faltered, unable to compete with the manned platforms of the era. The missiles were less accurate, more vulnerable, inflexible, and not necessarily cheaper than manned bombers. The Air Force's first major cruise missile program, the Snark "flying bomb" (an early version of an air-breathing cruise missile), struggled to meet accuracy and reliability requirements in testing. Famously, the system's high failure rate led the waters near the test range to be colloquially dubbed "Snark infested waters."[99] The Air Force's follow-on to the Snark, the Navaho, also struggled in testing, and ultimately Strategic Air Command abandoned the cruise missile program in favor of ballistic missiles.[100] Cruise missiles continued in the Air Force but primarily under the shepherding of Tactical Air Command, which invested in cruise missiles as the Air Force's tactical unmanned component, developing and deploying the Matador and Maice (nuclear) cruise missiles to Europe and Asia as well as smaller air-launched missiles and decoys.[101] These early cruise missile systems were designed to target air defenses as well as other conventional military targets that might have otherwise been difficult for manned fighters and bombers to reach. The Korean War cemented their utility and kept the systems a top priority for Air Force investment, despite the relative power disparity between Strategic Air Command and Tactical Air Command.[102] Tactical Air Command's nuclear cruise missiles were ultimately decommissioned in the late 1960s when the Army's IRBM, the Pershing II, took over the mission.[103]

For the Navy, early investment in cruise missiles represented not so much an embrace of the technology as a deep distrust of the alternative, ballistic missiles—which they saw as dangerous for the aircraft carrier fleet. This led to the Regulus, the Navy's ship- and submarine-launched cruise missile,[104] deployed from 1955 to 1964. A follow-on to the Regulus, Regulus II, was canceled in favor of the submarine-launched Polaris ballistic missile and for a short time cruise missiles stagnated within the Navy. However, while the US Navy was focused on ballistic missiles, the Soviet Navy had invested heavily in cruise missiles. The 1967 sinking of the Israeli destroyer *Elath* by Soviet antiship cruise missiles convinced the Navy that they needed to reinvest in cruise missiles. Accordingly, the Navy initiated the Harpoon antiship cruise missile program in 1969 and began deploying the system on aircraft, surface vessels, and submarines starting in 1977.[105] The system was relatively short ranged, but its flexible use across platforms turned the Harpoon into the base model for a series of cruise missiles over the next forty years.

At the same time that the Navy was exploring the Harpoon, submarine advocate Admiral Rickover made a push for a new longer-range submarine-launched cruise missile (the Submarine Tactical Antiship Weapons System, or STAWS). Rickover hoped that the new cruise missile would lead to investments in a new submarine platform. The Navy wasn't initially interested. However, in 1972 the United States signed SALT I, which limited the number of submarine-launched ballistic missiles but did not explicitly discuss cruise missiles. This opened the window for a new submarine-launched cruise missile that the Navy argued could be used as a low-cost option for both strategic and tactical weapons.[106] Then secretary of state Henry Kissinger and secretary of defense Bill Clements liked the idea of cruise missiles as a bargaining chip for the next round of arms control negotiations, and subsequently the Navy and the Air force were told to cooperate to develop an air-launched cruise missile (ALCM) and a submarine-launched cruise missile.[107]

The unwieldy partnership Clements foisted on the Air Force and the Navy had a perhaps unforeseen byproduct of convincing two services that were only marginally interested in cruise missiles (and the Air Force perhaps not at all) that they needed to compete for control over the technologies. The Navy quickly took the lead. Their version of the sea-launched cruise missile, the Tomahawk, immediately outpaced the Air Force's half-hearted ALCM development. By the end of the Cold War, the United States had developed multiple versions of the Navy's Tomahawk, capable of being launched on submarine, aerial, and surface platforms and designed to target both land and naval targets. For the Air Force, cruise missiles were a direct competitor to the long-range manned bomber, and throughout the 1960s and early '70s, the Air Force drug its heels on cruise missile investments—slowing down, in particular, research and development for guidance technology. This led to a damning congressional report in 1973 that accused the Air Force of sabotaging the subsonic cruise aircraft decoy (SCAD) missile program, asserting that:

> The Air Force has proceeded with this program solely as a decoy, notwithstanding the direction of the Congress. It is generally recognized that the Air Force has resisted pursuing SCAD with an armed warhead because of its possible use as a standoff launch missile. This application could jeopardize the B-1 program because it would be necessary to have bomber penetration if a standoff missile were available as a cheaper and more viable alternative.[108]

However, cruise missiles got a technological reinvention with the microprocessor. Whereas 1950s inertial guidance had an inaccuracy of .03 degrees per hour, by 1970 this inaccuracy was down to .005 degrees per hour. Further, whereas previously inertial systems' large size (three hundred pounds) had necessitated larger cruise missiles (which also limited which platforms could launch them), post-microprocessor inertial navigation systems were only twenty-nine pounds.[109] These smaller, more maneuverable munitions, in conjunction with new computer-enhanced terrain mapping capabilities, meant that the cruise missile of the 1970s was more survivable against Soviet surface-to-air missiles than many manned aircraft, which were not optimized to fly risky low-altitude missions. This directly threatened the Air Force's preferred program, the B-1, a low-altitude/high-speed bomber designed to negate the increasing lethality of Soviet surface-to-air missiles and fighter interceptors. The problem for the B-1 was that advances in the microprocessor had made missiles far more effective—both for shooting down aircraft (hence the need for a low-altitude/high-speed bomber) and for guiding to and destroying ground targets. The Air Force had previously argued that missiles' accuracy limitations made manned bombers a necessary element of the nuclear triad. But with new missiles that were far more precise, had much greater ranges, and could evade air defense systems sometimes with better success than manned alternatives, the Air Force struggled to convince Congress or the executive that the United States needed a new manned bomber.[110] In fact, Carter almost nixed the B-1 entirely, preferring a new "cruise-missile carrier" over the Air Force's manned bomber proposal.[111] In the end, the Air Force argued that cruise missiles' guidance technology was too new and could not be trusted to consistently make it to the target.[112]

It was perhaps the Navy that ultimately saved the Air Force cruise missile investments. Direct competition between Navy and Air Force programs artificially protected the systems from intraservice internecine attacks. Congress was so impressed by the Navy's early progress over the Air Force that it initially cut ALCM funding altogether, which galvanized the Air Force to invest in its ALCM program, fearful they would otherwise receive a "torpedo rammed up its bomb bay." The competition between the Navy and the Air Force, plus the complications of SALT II, led to an Air Force compromise, made partly to stave off Navy encroachment into strategic missions—equipping the B-52 with nuclear-tipped cruise missiles and settling for a much-reduced B-1 program.[113] It also (begrudgingly) developed a ground-launched cruise missile (which the Army wanted even less to do with), which was decommissioned in the early 1990s after the United States and Russia signed the Intermediate-Range Nuclear Forces Treaty.

The final footnote to this discussion is the Army's role in smart munitions during the Cold War, most notably in land mines and, to a lesser extent, missiles. After World War II, land mines became far more prevalent in ground combat. The United States developed antitank, antipersonnel, and bounding mines, which they used extensively during the Korean War, often deployed by aircraft. Mines were seen as a necessary and effective tool to combat large-wave offensives (such as those seen in Chosin) and led to the development of the Claymore mine, an above-ground command-detonated mine that continued to be used in Vietnam.[114] Indeed, land mines evolved rapidly and were used extensively in Vietnam, particularly as a way to interdict Viet Cong land movements. The Air Force deployed wide-area antipersonnel mines, dragontooth mines, and gravel mines designed to induce the greatest casualties without necessarily leading to death. The horrific nature of these mines, their high cost (it was estimated that the systems cost the United States between $800 million and $2 billion), and the limited evidence of effectiveness put political pressure on the United States to move away from land mines, and the systems were put on ice after Vietnam.[115]

When it came to missile development—a technology the Army initially controlled—the Army was largely marginalized, both by competing armed services and by little support from inside the Army. Early attempts to develop an Army version of the V-2, the Redstone missile system, were sidelined by Air Force bureaucratic maneuvering,[116] and even Army IRBM developments, for example, the Jupiter program, were gobbled up by the Air Force's missile general, Schriever.[117] The mainstay of the Army's missile development during this era, the Pershing missile system, replaced the Air Force Mace nuclear-tipped cruise missile in Germany in 1965, and versions of the Pershing I and II were deployed throughout Europe until the systems were eventually shelved at the end of the Cold War.

Where the Army did have success was missile defense—a version of missiles that resembled most closely the artillery missions already part and parcel of the Army identity. As early as 1954, the Army deployed the Nike Ajax surface-to-air missile system, which was deployed across the United States to defend against encroaching Soviet air threats. This system received an upgrade in 1958 with solid fuel that extended the range and altitude of the missiles. Later, the Nike Hercules was developed with the capability to launch missiles with nuclear warheads.[118] The Army also developed the more mobile surface-to-air missile system, the Hawk, which debuted in 1959 and received significant electronic upgrades in the early 1970s as the I-Hawk. The system was eventually replaced by the Patriot in 1994. Finally, the Army did attempt to develop antiballistic missile defenses in the 1960s, but the program was canceled in 1972 after the Anti-Ballistic Missile Treaty was signed.

Strategic Unmanned Platforms: Space's Dominance

Smarter munitions may have dominated United States' Cold War unmanned arsenal, but unmanned platforms ruled the high altitudes, making unmanned platforms key components of strategic intelligence. The high altitudes had always been a draw for military innovators looking for an uncontested domain to collect intelligence. Starting at the end of World

War II, the US Air Force experimented with B-17/B-29 drones as well as high-altitude unmanned balloons to collect indicators of a Soviet nuclear launch.[119] Ultimately, the effort (Project Mogul) was abandoned—it was too difficult to control the drones and the technology and logistical support needed to remotely maintain the balloons' altitude was too expensive. However, the United States went on to experiment with unmanned balloons and photography, launching hundreds of intelligence balloons over the Soviet Union during the 1950s as part of Projects Moby Dick and Genetrix. The efforts were largely unsuccessful—the balloons were easy to shoot down and in the end the United States only recaptured 40 of the 448 balloons launched. The photos the United States did garner from the project were also of little intelligence value; because the balloons could not be controlled, most of the pictures ended up being of fields and forests.[120] To put the final nail in the coffin for the balloon systems, they were far from covert and caused a bit of a diplomatic furor when the Soviets learned of the program.[121] After a 1958 disastrous attempt at a high-altitude unmanned balloon reconnaissance program that ended up crashing in Poland, Eisenhower ended the program and doubled down on investments in the U-2, a manned high-altitude reconnaissance aircraft.

Beyond the earth's stratosphere, there were early post–World War II studies into the feasibility of space reconnaissance. A 1954 RAND report even called for a "satellite reconnaissance vehicle" as a "vital strategic interest to the United States." However, satellite development remained a low priority for both the Air Force and the Navy until an exogenous Sputnik shock in 1957. At the time, the Navy had a Vanguard satellite program in development, but the program had been kept secret and given a tight budget, ostensibly so it wouldn't compete with investments in ballistic missiles.[122] It also didn't help that then secretary of defense Charles Wilson didn't believe in satellites and showed open ambivalence to the Soviet efforts. All of that changed when the Soviets launched Sputnik into orbit, an event that riveted US onlookers and caused the national security community to worry about a missile gap with implications for both intelligence and nuclear stability.[123]

The investment in rockets for ballistic missiles reopened the aperture for unmanned space reconnaissance (even if they sometimes competed with satellites for funding). Satellites were easier to remotely control (and operate) and less likely to be shot down than unmanned aircraft—all factors that made them the preferred option for reconnaissance over high-altitude manned and unmanned flights. Further, as more and more satellites were launched, the systems made persistent monitoring of strategic activities, like nuclear weapons development, more feasible. All these factors made space platforms ideal tools for early warning of nuclear launches (a key element of US nuclear second-strike survivability) as well as the monitoring of missile facilities and launch sites that was necessary to validate arms control agreements of the latter half of the Cold War. As John Deutch, former undersecretary of defense for acquisition and technology and director of the Central Intelligence Agency, reminisced, "I was very heavily influenced by satellite reconnaissance—the first large case of unmanned systems where they got rid of the SR-71 and other manned systems trying to do reconnaissance against the Soviet Union. I was of the view that satellite reconnaissance instead of manned reconnaissance was important for arms control reasons."[124]

In contrast to space reconnaissance, the Air Force's unmanned platforms—which were primarily designed for airborne reconnaissance—never became a significant element of US strategic forces. This trajectory was not a foregone conclusion. Congress and senior military decision-makers were thirsty for unmanned options that would allow the United States better covert warning of Soviet nuclear developments—especially without the risk of losing pilots to increasingly capable surface-to-air missiles. Despite the political desire for these systems, they never received significant support outside of "black" (covert) programs or highly classified Strategic Air Command unmanned reconnaissance sorties—mainly used to plug gaps in coverage of Chinese nuclear development, created by scarce and expensive satellites otherwise devoted to USSR targets.[125] In the end, even these unmanned aircraft systems that Strategic Air Command used for long-range reconnaissance were often high cost and technologically unreliable—two characteristics that even deep pockets were not willing to support, especially with the comparative advantage of satellites.[126]

This focus on satellites for strategic missions led to a proliferation of satellites capable of collecting signals intelligence, taking electro-optical images, and transmitting executive orders about nuclear use. Their strategic and highly classified nature also kept them relatively immune from service fights, as satellites were deemed "national" assets that prioritized presidential and national intelligence priorities. Space represents one of the few times that the Air Force was not able to control a mission it had initially developed. In 1960 a special commission set up by the Eisenhower administration determined that the Air Force had poorly run early satellite development efforts, known as WS-117L and Project Corona. The commission was also concerned that overt uses of space for military missions might create a space race with the Soviets. Accordingly, the administration placed satellite reconnaissance under the control of a civilian agency, the National Reconnaissance Office, and created covert satellite reconnaissance designations. From that point on, the Air Force's only space missions would be in military support missions (communications, early warning, weather, navigation, etc.)—mostly connected to strategic missions.[127] Even those missions would later be divided between the services, with the Navy in charge of navigation satellites, the Army in charge of communication, and the Air Force the lead on nuclear early warning (the Defense Support Program satellite constellation and Vela Hotel satellites).[128] This rather arbitrary division of responsibility within space decreased service fights over space, but it also led to a lower priority in service budgets—especially when the resources served joint purposes. For example, investments in global positioning systems (GPS) stalled in the 1980s as the services focused on big-ticket platforms within their budgets. It was not until the Gulf War that the Air Force doubled down on GPS investments (and this was after the Air Force tried multiple times to cut the GPS program).[129] Together, the limited amount of satellite resources available to the services to support their core missions and civilian control over the resources created an impetus for the next big trajectory in unmanned experimentation—the tactical unmanned aircraft.

Tactical Experimentation: Korea, Vietnam, the Rise of Unmanned Aircraft, and Tactical Smart Munitions

While the United States was doubling down on ballistic missiles, nuclear-tipped cruise missiles, and satellites to deter strategic fights with the Soviet Union, two conflicts carved out space for tactical and operational experimentation with unmanned systems. The Korean War in 1950 showcased the need for conventional air-breathing guided missiles, while two decades later Vietnam was a proving ground for unmanned aerial platforms and munitions. After largely ignoring guided bombs after World War II, the need for more accurate combat air support during the Korean War led the Air Force to dust off a radio-controlled bomb designed at the tail end of World War II, the RAZON VB-3, and subsequently modified the TARZON, a twelve-thousand-pound bomb, with RAZON's radio guidance.[130] The Korean War also drove investments away from the Navy's preferred antisubmarine missions and spurred investment in air and surface missiles.[131] In particular, the Navy ship-based Terrier[132] and Sparrow air missile systems as well as the Army Nike ground-based missile system were put on the fast track for development because of the Korean War.[133]

Between the Korean and the Vietnam War, autonomy for tactical missions received less attention than strategic counterparts. A focus on nuclear weapons, especially within the Air Force, left few advocates for nonnuclear guided munitions (or the platforms to deliver them).[134] However, in the late 1960s, the United States found itself embroiled in a new "limited" war in Vietnam. As the war expanded and at the same time became less politically tenable, political decision-makers and campaign commanders sought unmanned options to reduce US casualty counts and provide much-needed (and otherwise dangerous) combat reconnaissance, including an ambitious project pioneered by Westmoreland to create an "invisible fence" of unmanned surveillance in Vietnam that used mines, acoustic and seismic sensors, and manned relays to detect and interdict attacks.[135] In practice, the invisible fence struggled to perform as envisioned; sensors were finicky, manned relay aircraft were costly to maintain and sustain, and the "real time" intelligence often lagged too far behind actual events to successfully respond. However, the experimentation led to one of the first machine integrations of information within a real-time display, a harbinger of later machine learning–enabled operating pictures.

The contentious domestic political situation created by Vietnam was exacerbated by the advent of Soviet surface-to-air missiles, which made American pilots vulnerable to shoot down and capture. This confluence of tactical and political issues empowered the Air Force's Tactical Air Command, which had played a secondary role behind the more powerful Strategic Air Command, to nudge autonomous systems away from their previous strategic focus.[136] The first shift was in the early experimentation of precision-guided munitions. Tactical Air Command repurposed a radar-bombing method called Skyspot from Strategic Air Command, which had been using the method in testing, to guide otherwise dumb bombs from B-52s in what was one of the early innovations in precision-guided munitions.[137] The use of offboard and onboard radar sensors (e.g. within new long-range navigation radars in F-4s) to guide bombs set the stage for more autonomous munitions, with longer (and smarter) hand-offs, all while increasing the effectiveness of the manned platforms that dropped the bombs. The Navy's early successful investments in autonomous munitions during the 1950s and '60s meant that early guided munitions innovations in Vietnam, including the Bullpup, Shrike, and Walleye bombs, were primarily naval. However, as the conflict progressed, the Air Force adapted Navy innovations and developed its own smart munitions including the homing bomb system, electro-optical-guided GBU-15, rocket-propelled Maverick, and Paveway/knife/spike laser-guided bomb.[138] Even the AIM-7 Sparrow air-to-air missile, despite a slow start and dubious initial effectiveness, got a jumpstart during Vietnam.[139] All these systems paved the way for the precision-guided munition revolution that was looming on the horizon—including the AMRAAM air-to-air missile, the semiactive radar-killing HARM, and the Navy's AIM-54 air-to-air Phoenix missile.[140]

Tactical Air Command also convinced Strategic Air Command to reapportion the Lightning Bug, an unmanned reconnaissance system that Strategic Air Command was considering as a replacement for manned U-2 flights over China, from a strategic reconnaissance mission to a tactical support mission. The Lightning Bug was a derivative of a joint project to develop a target drone, the Ryan Firebee (BQM-34A). The system, which began development in 1958, received higher priority after the Cuban Missile Crisis and the downing of the U-2 but largely languished through the 1960s, used only sparingly for reconnaissance flights over China. Vietnam took the system from experiment to operations. The Lightning Bug and later Buffalo Hunter drones flew over four thousand sorties in Vietnam, serving as bait for manned missions, taking photos of Vietnamese surface-to-air missiles and prison camps, conducting poor-weather battle damage assessments, and dropping propaganda leaflets.[141] The drones were useful complements to manned bombing runs; because they were small and flew at low altitudes and high speeds, they were hard to detect. Further, the systems were supported by electronic countermeasures and remotely controlled, allowing pilots to maneuver and evade threats.[142] It wasn't full autonomy, but it was a model for remote-control unmanned piloting that adapted well to the tactical air fight.

The Navy's investment in unmanned aerial systems provides an interesting counterpoint to the Air Force. The Cold War Navy invested very little in unmanned platforms—whether at sea, underwater, or in the air. An unmanned version of the F-6 Hellcat, laden with explosives and controlled via radio by an accompanying Skyraider manned aircraft, was used to target North Korean bridges and tunnels.[143] The unmanned Hellcat achieved minor success, but the systems were primarily used as target drones to test air-to-air missiles. In fact, an unfortunate incident in California in which operators lost control of a target drone over a residential area led to over a thousand burned acres in the California desert.[144] Only after Vietnam (and the loss of Navy pilots in missions in Lebanon) did the Navy invest in the Israeli Pioneer. The Pioneer unmanned aerial vehicle went on to see use in the Gulf War (mostly by the Marines), though the Navy ultimately abandoned the Pioneer program in the 1990s.[145]

Finally, the Army—bequeathed only short-range and primarily defensive missiles or small unmanned aerial systems in the unmanned service food fight—was almost completely left out of these unmanned trajectories. At the same time that the space and air domains were exploding with unmanned development, ground unmanned systems made little advancement and almost no integration into operations outside of reconnaissance or mine removal.[146] Part of this was due to the technological challenge of creating unmanned ground systems that could navigate terrain. Microprocessors could decrease the space required to conduct computing, but the computers of the 1960s and '70s were still unable to tackle some of the more complicated machine learning tasks of moving through cluttered terrain. Therefore, while the Army and Marines invested in a series of experimental efforts with unmanned robotics and ground systems, the technology required to effectively control these systems in battle was not yet mature enough to bring these technologies into combat. The Army's primary investment in unmanned systems during this period occurred in Vietnam, where the Army turned to fixed unmanned reconnaissance sensors (coupled with aerial systems) to help mitigate tactical losses on the ground. After Vietnam, these Army-initiated unmanned listening posts and partner aerial vehicles programs languished until they were either canceled or co-opted by the Air Force. Toward the end of the Cold War, unmanned sensors and tactical drones were highlighted as part of the Army's AirLand Battle Corps 86 acquisition program, but these systems wouldn't transition from dream to reality until after the fall of the Berlin Wall.[147]

While this initial foray into the integration of tactical unmanned systems was ultimately abandoned by the Air Force after the war, the lessons of unmanned experimentation in Vietnam were foundational to the investment in and adoption of unmanned aerial systems in the subsequent decades (and the dominance of this trajectory in contemporary unmanned systems investments). Experimental uses of unmanned aerial vehicles for electronic warfare countermeasures and suppression of enemy air defenses—first in Vietnam and then by Israel in the Arab-Israeli wars—demonstrated new mission sets for unmanned aerial systems and influenced investments in unmanned systems within the other services.[148]

As the United States came out of the Cold War, the air and space domains dominated unmanned innovation. However, within the air domain, there was significant competition for investment. Strategic missile systems and satellites initially dominated the inventory, complementing each other to enable strike and intelligence missions. However, the confluence of technological development (i.e., the microprocessor), a politically fraught and bloody war in Vietnam, and the bureaucratic rise of organizations like Tactical Air Command opened the aperture for unmanned aerial vehicles, guided tactical munitions, and tactical uses of unmanned systems—especially as a manned substitute for intelligence collection in dangerous scenarios. In the sea, while unmanned platforms and early leaders like mines and torpedoes saw little development, cruise and ballistic missiles thrived. Finally, on the ground, manned systems dominated across mission sets as technological challenges, bureaucratic losses, high cost, and questionable ethics of land mines put unmanned technology at a comparative disadvantage to manned alternatives.

The Gulf War and Beyond: The End of History and the Beginning of Our Story

If the microprocessor reduced the physical computing limitations of unmanned systems, the advent of the internet and digital technologies unlocked the virtual computing boundaries of unmanned systems. For the first time in history, these systems could be reliably remotely controlled and sensors could transmit large troths of information in real time. The ability to collect, store, and eventually transmit large packets of digital information elevated the importance of unmanned aerial vehicles (which had previously lagged behind missiles and space systems) and allowed for both risk mitigation and persistent intelligence functions. It also secured unmanned systems' role as tactical complements to manned weapons within conventional campaigns. This move was no doubt spurred by the dissipation of the Cold War and the proliferation of smaller-scale conflicts. Accordingly, the mass firepower of strategic ballistic missiles (and the dominance of a now-defunct Strategic Air Command) gave way to a lower-yield but more practical unmanned system—the cruise missile. Campaigns in Iraq, Somalia, and Bosnia became proving grounds for these new digitally enabled unmanned systems. Their ability to lean on the networks of a modern military, strike from increasingly long distances, and provide risk-free, persistent, and immediate intelligence made unmanned systems a key part of strategic success for the US military. By the unveiling of the Department of Defense (DoD)'s first unmanned systems roadmap in 2007, unmanned systems had officially made the transition from novelty, experimentation, and strategic munitions to widespread proliferation of tactical missiles alongside increasingly automated and fielded unmanned air platforms.

But the proliferation of these systems did not happen equally along all mission sets, domains, and services. Instead, some trajectories skyrocketed while others stagnated in largely the same status as their Cold War heyday. Nowhere was this dichotomy more apparent than in the air domain, where, after years of failed attempts at unmanned aircraft programs, unmanned aircraft systems began to compete for resources with missile systems that had taken most of the unmanned budget throughout the Cold War. The transition away from strategic missiles to unmanned aerial vehicles was ushered by the downfall of the Cold War and the demise of Strategic Air Command. The 1990s saw a unipolar world with no looming peer competitors to race for strategic dominance. Ballistic missiles, which had received sizable investments in the Cold War as a major linchpin of the US deterrence strategy against the USSR, sat in their silos with little more than routine maintenance. Meanwhile, the strategic air forces of the United States focused their efforts from nuclear deterrence toward manned global strike (which was capable of a dual nuclear-conventional strike capability), undergirded by a network of satellites that would continue to dominate overall unmanned investment in intelligence missions. However, the high cost of satellite reconnaissance (which made them a national asset instead of a battlefield commander's resource) and the lag between satellite imagery and real-time imagery and videos created a space and desire for tactical unmanned aerial vehicles able to provide persistent coverage over targets, increasingly with real-time video download capability.

While the United States might not have had a looming peer competitor after the Cold War, the US military was still embroiled in conflicts. Over the course of the 1990s, the United States opted into a series of lower-intensity conflicts in which policy makers (and defense practitioners) sought to achieve victory with minimal human cost. This led to a focus on cruise missiles and precision munitions. First, the cruise missile and other long-range missile technologies, which had begun to gain prominence after the invention of the microprocessor and the demonstrated success of long-range attacks during the Arab-Israeli conflicts of the 1970s, took off after the first Gulf War. These long-range attack missiles were equipped with new digital targeting capabilities, making them more accurate (and maneuverable) than previous missile systems. The long range at which they could be fired allowed the United States to implement a shock-and-awe doctrine for significant campaigns like Desert Storm and facilitated low-political-risk engagements against terrorists, drug lords, and genocide leaders throughout the 1990s and into the early 2000 conflicts in both Afghanistan and Iraq. Meanwhile, investments in precision-guided munitions, including active and semiactive air-to-air and air-to-ground missiles, laser-guided bombs (the Paveway series), and GPS-guided bombs dramatically increased the effectiveness and range of manned airpower.

These low-stakes conflicts provided the initial impetus for research, investment, and early experimentation with what would become the MQ-1 Predator unmanned aerial vehicle.[149] Further, the advent of digital wireless network technologies that could transmit both within line of sight and (with satellite relay) over the horizon made it possible for the Predator tactical unmanned reconnaissance mission to transmit videos to field and combatant commanders as well as stateside command staffs—all doubling down on situational awareness to avoid embarrassing military failures. While the Predator debuted in the 1990s, it cemented its breakout role in modern conflict after 9/11. Leading up to 9/11, the Predator was a niche project, funded almost completely out of hide by a defense company that believed in the revolutionary role of these unmanned technologies.[150] But after 9/11, the Predator moved out of its fringe role and vaulted into one of the most iconic weapons platforms of the global war on terror. Perhaps most importantly, 9/11 created the impetus to arm what had been a reconnaissance-only unmanned system, thereby introducing to the battlefield one of the first armed unmanned platforms (there had been many other unmanned munitions). As then secretary of the Air Force James Roche recounted, there was a lot of resistance to arming the Predator before 9/11, but then "two buildings fell over"[151] and suddenly arming the Predator didn't seem nearly as risky or revolutionary as it had previously.

Not surprisingly, unmanned investment post-9/11 was heavily focused on platforms within the air domain. The MQ-1 and its successor the MQ-9 became integral to campaigns in Iraq and Afghanistan, both for persistent reconnaissance and (increasingly) strike missions. Further, a long-range reconnaissance air platform, the Global Hawk (meant to replace the manned U-2), also debuted in the decade after 9/11 and began to provide longer-range strategic coverage of near-peer competitors. While the Predator, Reaper, and Global Hawk were owned and operated by the Air Force, the Army also invested heavily in (generally smaller) unmanned systems for tactical reconnaissance. Unlike the Air Force, which treated their unmanned air platforms as higher-command strategic assets, the Army integrated their unmanned air systems within their units as part of regular army units. Meanwhile, the Navy embarked on experimentation efforts with unmanned aircraft on carriers, though they lagged behind both the Air Force and the Army in their integration of unmanned air systems in combat.

Unmanned platforms in the maritime domain, in general, did not get the same kind of exponential increase in support or investment as the air domain. Unmanned underwater vehicles and unmanned surface vehicles, while often touted in research and development documents during the early 2000s as examples of transformative technologies, largely stagnated in research and development. The Navy focused on the large manned shipbuilding initiatives that took most of their research and procurement budgets. Meanwhile, they relied on existing manned air platforms and missile inventories to support conflicts in Iraq and Afghanistan. Their land component, the Marines, pushed harder for unmanned tactical reconnaissance (similar to Army initiatives), but this effort didn't translate into significant changes in overall unmanned naval investments.

On the ground, the Army continued to invest in air defense missile systems, including the Patriot surface-to-air missile system, as well as increasingly precise rocket systems including the Multiple Launcher Rocket System (MLRS) and the M142 High Mobility Artillery Rocket System (HIMARS). It also initiated a significant and comprehensive transformation program, the Future Combat System. The future combat system sought to network Army systems together and team both manned and unmanned ground vehicles with unmanned sensors and unmanned aerial vehicles to create an entirely new concept of operations. Accordingly, the Army invested in new technologies to develop unmanned ground vehicles and, in the field, experimented with and implemented robotic systems to combat and disarm improvised explosive devices, which had become a prevalent threat during campaigns in both Iraq and Afghanistan. The program ultimately failed, but some of the initiatives in tactical unmanned aerial vehicles and unmanned robotics for disarming explosives normalized and became increasingly pro-lific elements of warfare in the Middle East.

All of these efforts toward unmanned investment after 9/11 were helped by significant congressional support. For instance, in the defense budget of fiscal year 2001, Congress tasked the DoD with greater investment in unmanned systems, requiring that by 2010, "one third of the aircraft in the operational deep strike force should be unmanned," and "by 2015, one third of the Army's Future Combat Systems operational ground combat vehicles should be unmanned."[152] Large congressional support, big budgets, and an opportune combat environment created ideal conditions for unmanned expansion in the DoD. By 2007 and the publication of the DoD's first Unmanned Aerial Systems Roadmap, a "coalition Unmanned Aircraft Systems (UASs), exclusive of hand-launched systems, had flown almost 400,000 flight hours in support of Operations Enduring Freedom and Iraqi Freedom, Unmanned Ground Vehicles (UGVs) had responded to over 11,000 Improvised Explosive Device (IED) situations, and Unmanned Maritime Systems (UMSs) had provided security to ports."[153] While the vast majority of unmanned systems continued to be focused on reconnaissance, they were increasingly being used in strike capacity and—perhaps most notably for the contemporary push for unmanned systems—were now viable not just as munitions or as strategic resources but also remotely controlled platforms at the operational and tactical level of warfare.

Contemporary Unmanned Systems

This leads us to contemporary unmanned spending. The fiscal year 2020 National Defense Authorization Act (NDAA) authorized over $10 billion for unmanned systems, including $3.1 billion for unmanned platforms; $6 billion for tactical, strategic, and intercontinental missiles; and $1.3 billion for space systems. And while this is probably only a portion of the total outlays for some of these classified systems, cumulatively they made up 1.5% of the total fiscal year 2020 budget.[154]

Unmanned systems got their start in the air, and so it is no surprise that even accounting for space, the air has seen the greatest proliferation and maturation of unmanned systems in both platforms and missiles. Over 65% of the 2020 budget appropriation was dedicated to unmanned systems, with a little over half of that devoted to unmanned aerial systems. The air domain includes what are probably the best known of the unmanned contemporary inventory, the MQ-1 Predator (now retired) and MQ-9 Reaper. While both of these famous unmanned systems are capable of strike missions, the unmanned aerial platform inventory has expanded considerably to a series of intelligence, surveillance, and reconnaissance aircraft in the Army, Air Force, Navy, and Marines (RQ-11, RQ-7, RQ-4, RQ-170, RQ-20, RQ-12, etc.). Further, the Navy's unmanned carrier-capable aircraft, the X-47B, introduces a logistical (refueling) capability to the unmanned air inventory. Almost all of these platforms require remote control, either from line of sight or via satellite across the horizon. While most of the strike platforms are of comparable size to manned aircraft, many of the smaller airborne intelligence, surveillance, and reconnaissance platforms are small enough to be operated by infantry or special operations units. Missiles also continue to be a large portion of the unmanned inventory, especially air-to-ground missiles, air-to-air missiles, and ICBMs.

While the air domain has seen the most extensive proliferation of unmanned technology, the naval domain also includes investments in underwater drones, including systems that specialize in mine countermeasures, identification, or disposal (Sea Fox, Barracuda, Mark 18, etc.), as well as underwater and surface intelligence (Sea Stalker, Sea Maverick, etc.). A large part of the Navy's investment in unmanned technology is in munitions and, in particular, updates to submarine-launched ballistic missiles. Further, while not in the 2020 NDAA, later budgets called for significant investments in unmanned surface ships—part of the Navy's larger strategy to increase the size of the Navy by augmenting manned platforms with unmanned.[155]

Finally, the ground domain continues to lag behind investment in all the other domains. While the Army has invested in unmanned aerial systems for tactical reconnaissance, it has not paid the same amount of interest to or invested similar resources in the ground domain. Investments in unmanned ground systems have been mostly in explosive ordnance disposal as well as recent investments in testing unmanned combat vehicles.

The air domain may lead in unmanned appropriations, but in the fiscal year 2020 NDAA the Navy received the largest amount of unmanned funding—mostly because of significant investments in new programs to develop both unmanned aerial platforms and unmanned underwater vehicles. The Army lagged behind both the Air Force and the Navy, and most of this investment continued to be in the air (and not the ground) domain as well as (increasingly long-range) surface missiles. The Air Force, with years of research and fielding of unmanned systems, focused its budget on procurement (buying) and maintenance of programs, while the Navy, which lagged behind the Air Force for many of these initial programs of record, devoted most of its resources to development (Research, Development, Test and Engineering, RDT&E). Similarly, Army efforts outside of the air domain were mostly focused on RDT&E with procurement and maintenance primarily of its unmanned aerial systems used mostly for reconnaissance and limited strike. Finally, while the vast majority of unmanned system investment was in strike capabilities, this is skewed heavily by missiles. Unmanned platforms were still predominantly designed for intelligence, surveillance, and reconnaissance missions.

The fiscal year 2020 and the next year fiscal year 2021 budget also introduced an important trajectory for unmanned systems that extended beyond missions, services, or domains. This is the investment in autonomy and artificial intelligence beyond munitions or platforms. As Konaev et al. summarize, "The FY2021 US defense budget request allocates $1.7 billion to autonomy to enhance 'speed of maneuver and lethality in contested environments' and the development of 'human/machine teaming,' as well as $800 million to AI."[156] The new budget represented increased investment in networks, technology, and organizations required to use autonomy across systems. Further, by breaking out autonomy from artificial intelligence, it also showed a focus on the algorithms and data that underlie any artificial intelligence required to increasingly transition unmanned systems from remotely controlled systems to complete autonomy. This focus on autonomy over simply taking the man out of a system and the transition in vocabulary from "unmanned" to "autonomous" cemented a large shift in the trajectory of these systems.

Conclusion

This short historical outlay shows that the subset of systems that dominate "unmanned" narratives is only a small proportion of autonomous weapons and platforms within the US inventory. Unmanned systems today represent a greater percentage of the military force, perform a wider array of missions, and operate under a broader spectrum of control than could have been imagined even fifty years ago. They are of every shape and size, exist across domains, and are of increasing levels of automation. Unmanned systems range from small, off-the-shelf, remotely controlled microdrones to remotely controlled strike-capable aircraft. On the ground, they can be as small as trunk-size robots or as large as unmanned transport vehicles. And in the sea, unmanned systems operate underwater and on the surface, sometimes as unsophisticated as dumb mines and other times as complex as unmanned surface vehicles. They include missiles, robots, aircraft, ships, submarines, and satellites—although the vast majority of contemporary focus has been on unmanned platforms, unmanned munitions and sensors have historically played a larger role.

Further, as Table 1.1 demonstrates, the story of how we got to contemporary US unmanned systems demonstrates how patterns of unmanned development emerged even prior to the investment that led to today's systems. First, the air and sea domains have always had more success with unmanned experimentation than the ground. This is partly a technological phenomenon of control and how the development of wireless communication allowed for (pre-microprocessor) imprecise control of unmanned systems. Imprecise control was enough to drive innovation for air and (to a lesser extent) naval systems; it was not enough to prove the utility of unmanned ground systems, which needed to deal with a far more cluttered environment. Control was a major consideration for where unmanned systems succeeded and where they were ultimately discarded. Unmanned missions that needed control—for example, strike missions near civilians or friendly troops—were quickly abandoned for those, like reconnaissance or countermines, that could be indiscriminate and persistent. In general, though, across domains the trade-off between remote-controlled and autonomous technologies meant that unmanned platforms (which needed to return to sender) were often abandoned for their far more cost-effective autonomous unmanned munitions. If it was a choice between an unmanned surface ship or a torpedo, an unmanned strategic intelligence, or a missile or an unmanned tank or a surface torpedo, leading up to 9/11, the munition always won out over the platform. It was simply less technically difficult to develop, maintain, and control.

The trajectory of unmanned systems after 9/11 demonstrates a major divergence from previous concerns about control. Expensive, remotely controlled unmanned aerial platforms for both reconnaissance and strike missions dominated US military investments. These systems saw significant use in Iraq and Afghanistan, ultimately proving the utility of remotely controlled

Table 1.1 US Unmanned Trajectory

Up to World War I	Dominated by naval innovations: torpedo, mine—mostly unguided or crude guidance. Early attempts at unmanned balloons for signaling, artillery led to Kettering Bug unmanned aircraft in World War I.
Interwar–World War II	Experimentation with unmanned for training: surface naval vessel, unmanned aerial vehicles. Continued development (albeit slowed) for torpedoes and mines. Experiments with unmanned aerial vehicles for strike during World War II. Late World War II development of long-range rockets/missiles.
Early Cold War	Focus on nuclear armed ballistic missiles (Air Force/Schriever, Navy/Rickover), space satellites for reconnaissance after Sputnik (primarily civilian control), torpedo/mine development stagnates while Air Force/Navy/Army half-heartedly explore nuclear cruise missiles. Focus on strategic systems.
Vietnam–End of Cold War	Tactical innovation and unmanned for force protection (Westmoreland and the invisible fence)—Vietnam leads to unmanned aerial vehicles for bait, battle reconnaissance (Lightning Bug), renewed interest in guided munitions. Arms control agreements and air defense improvements push cruise missiles over ballistic missiles.
End of Cold War–9/11	Focus on Powell doctrine/force protection, rise of technological narratives, decrease in military budgets. Cruise missiles prioritized, precision-guided munitions and space-based navigation become prominent; Gulf War cements investments in precision strike. Unmanned aircraft primarily novelty.
9/11 and Beyond	Rise of armed unmanned aerial vehicles, focus on permissive air environments, proliferation of space-based support, reconnaissance and rise of Space Force. Nuclear ballistic missiles and conventional strike stagnate. Block munition upgrades, limited new inventory.

unmanned systems in modern combat. Moving into 2020, the success of these systems animated interest across the services in unmanned aerial systems. All of the services invested heavily in unmanned aerial systems for intelligence as well as strike and, to a lesser extent, logistics, with the Army and the Navy playing catch-up in research and development of these platforms, while the Air Force outlays most of its resources to improving and maintaining existing remotely controlled unmanned aerial vehicles. Despite the rise in unmanned aerial platforms, increasingly long-range munitions continue to receive significant (even if not priority) investment, demonstrating the enduring tension between munitions and platforms in unmanned systems.

How we got to this current outlay of unmanned systems is a story of technology, cost, and war. But these equations of capabilities, expenses, and conflict weren't simple, and decisions about unmanned systems quickly became intertwined with the strategies and identities of organizations that owned those budgets. Untangling the knot between capabilities, beliefs, and identities to explain how we got from the Cold War to today will be the focus of the rest of this book.

Notes

1. Lawrence Freedman, *The Future of War: A History* (New York: Public Affairs, 2017), xvii.
2. For a good discussion of different ranges of autonomy, see Scharre, *Army of None*.
3. Kimon P. Valavanis and Michail Kontitsis, "A Historical Perspective on Unmanned Aerial Vehicles," in *Advances in Unmanned Aerial Vehicles*, ed. Kimon P. Valavanis (Dordrecht: Springer, 2007), 15–46; Norman Youngblood, *The Development of Mine Warfare: A Most Murderous and Barbarous Conduct* (Westport, CT: Praeger Security International, 2006), 4.
4. Youngblood, *The Development of Mine Warfare*, 5–7.
5. Dennis Hart Mahan, *A Complete Treatise on Field Fortification, with the General Outlines of the Principles Regulating the Arrangement, the Attack, and the Defense of Permanent Works* (New York: Greenwood Press, 1968), 75.
6. Youngblood, *The Development of Mine Warfare*, 26.
7. Ibid., 26–38.
8. James Saint Claire Morton, *Memoir of an American Fortification* (Washington, DC: William A. Harris, 1959), 496–497.
9. Youngblood, *The Development of Mine Warfare*, 41–44.
10. Ibid., 43.
11. As quoted in ibid., 57.
12. Howard S. Levie, *Mine Warfare at Sea* (New York: Martinus Nijhoff Publishers, 1992).
13. Francis Hopkinson, *The Battle of the Kegs* (Boston: Oakwood Press, 1866).
14. Youngblood, *The Development of Mine Warfare*, 13.
15. Robert Fulton, *Torpedo War and Submarine Explosions* (Chicago: Swallow Press, 197 [reprint]).
16. Levie, *Mine Warfare at Sea*; Thomas Wildenberg and Norman Polmar, *Ship Killers: A History of the American Torpedo* (Annapolis, MD: Naval Institute Press, 2010), 7–17; Roger Branfill-Cook, *Torpedo: The Complete History of the World's Most Revolutionary Naval Weapon*

(Annapolis, MD: Naval Institute Press, 2014); Anthony Newpower, *Iron Men and Tin Fish: The Race to Build a Better Torpedo during World War II* (Westport, CT: Praeger, 2006).

17. Youngblood, *The Development of Mine Warfare*, 45.
18. Gregory Hartmann, *Mine Warfare History and Technology* (Silver Spring: Naval Surface Weapons Center, White Oak Laboratory, 1975), 10–12.
19. H. R. Everett, *Unmanned Systems of World War I and II* (Cambridge, MA: MIT Press, 2015).
20. Wildenberg and Polmar, *Ship Killers*, 13.
21. Ibid., 12.
22. Ibid., 18, 26–27.
23. Katherine C. Epstein, *Torpedo: Inventing the Military-Industrial Complex in the United States and Great Britain* (Cambridge, MA: Harvard University Press, 2014).
24. Ibid., 10.
25. Chief of the Bureau [Ordnance] to President of the Torpedo Board, in *Report of the [Senate] Select Committee on Ordnance and Warships* (Washington, DC: GPO, 1886), 151–153.
26. Epstein, *Torpedo*, 218.
27. Christopher H. Sterling, *Military Communications: From Ancient Times to the 21st Century* (Santa Barbara, CA: ABC-CLIO, 2008).
28. Charles A. Ziegler, "Weapons Development in Context: The Case of the World War I Balloon Bomber," *Technology and Culture* 35, no. 4 (1994): 750–767.
29. Lee Kennett, *The First Air War: 1914–1918* (New York: Simon and Schuster, 1999); Steven D. Culpepper, *Balloons of the Civil War* (Auckland, NZ: Pickle Partners Publishing, 2014).
30. Laurence R. Newcome, *Unmanned Aviation: A Brief History of Unmanned Aerial Vehicles* (Reston, VA: American Institute of Aeronautics and Astronautics, 2004); Kimon P. Valavanis and George J. Vachtsevanos, eds., *Handbook of Unmanned Aerial Vehicles*, vol. 1 (Dordrecht, Netherlands: Springer, 2015); John F. Keane and Stephen S. Carr, "A Brief History of Early Unmanned Aircraft," *Johns Hopkins APL Technical Digest* 32, no. 3 (2013): 558–571; Sarah Elizabeth Kreps, *Drones: What Everyone Needs to Know* (Oxford: Oxford University Press, 2016); John Kaag and Sarah Kreps, *Drone Warfare* (Hoboken, NJ: John Wiley & Sons, 2014); Kennett, *The First Air War*.
31. Keane and Carr, "A Brief History," 558–571.
32. Steven J. Zaloga, *Unmanned Aerial Vehicles: Robotic Air Warfare 1917–2007* (London: Bloomsbury Publishing, 2011); Everett, *Unmanned Systems of World War I and II*.
33. Everett, *Unmanned Systems of World War I and II*, 77.
34. Joseph V. Gluth, "Is the Navy's Mine Warfare Posture Bankrupt?" (thesis, US Naval War College, 1991); Levie, *Mine Warfare at Sea*.
35. Youngblood, *The Development of Mine Warfare*, 68.
36. Ibid.
37. Wildenberg and Polmar, *Ship Killers*, 43.
38. Ibid., 562.
39. Newpower, *Iron Men and Tin Fish*.
40. Neufeld, *The Development of Ballistic Missiles in the United States Air Force, 1945–1960*, 8–9.
41. Everett, *Unmanned Systems of World War I and II*, 105.
42. Ibid., 424.
43. William King. *Torpedoes: Their Invention and Use from the First Application to the Art of War to the Present Time* (Washington, DC: n.p., 1866), 86–87.
44. *Complete U.S. Infantry Guide* (Philadelphia: J. B. Lippincott, 1917), 1237.
45. Everett, *Unmanned Systems of World War I and II*, 425.
46. John H. Hammond Jr., Statements before the Hammond Torpedo Board, in Hearings before Subcommittee of House Committee on Appropriations, 65th Congress, Third Session, regarding torpedoes for coastal defense, October 3–9, 1918.
47. Hammond, Statements before the Hammond Torpedo Board.
48. Kenneth Werrell, *The Evolution of the Cruise Missile* (Maxwell, AL: Air University Press, 1985).
49. Naval War College, "Tactical Problem IX: Class of June 1919" (Newport, CT: Naval War College, June 3, 1920). Copy held in John Hays Hammond Papers, MS 863, Box 5, Yale University Library, New Haven, CT.
50. Ibid.
51. Volker Bertram, "Unmanned Surface Vehicles—A Survey," Published January 2008, http://citeseerx.ist.psu.edu/viewdoc/download?doi=10.1.1.462.1894&rep=rep1&type=pdf.

52. Werrell, *The Evolution of the Cruise Missile*, 25.
53. Everett, *Unmanned Systems of World War I and II*.
54. Robert Gannon, *Hellions of the Deep: The Development of American Torpedoes in World War II* (University Park: Pennsylvania State Press, 1996), 33.
55. Wildenberg and Polmar, *Ship Killers*, 81; Newpower, *Iron Men and Tin Fish*; Gannon, *Hellions of the Deep*.
56. Wildenberg and Polmar, *Ship Killers*, 91.
57. Ibid., 97–100.
58. Youngblood, *The Development of Mine Warfare*, 110–141.
59. Everett, *Unmanned Systems of World War I and II*.
60. Keane and Carr, "A Brief History," 558–571.
61. Bishane A. Whitmore, *Evolution of Unmanned Aerial Warfare: A Historical Look at Remote Airpower—A Case Study in Innovation* (Fort Leavenworth, KS: US Army Command and General Staff College, 2016).
62. John L. Frisbee, "Project Aphrodite," *Air Force Magazine* 57 (1997); Werrell, *The Evolution of the Cruise Missile*, 32.
63. Paul Gillespie, *Weapons of Choice: The Development of Precision Guided Munitions* (Tuscaloosa: University of Alabama Press, 2006), 29.
64. Alfred F. Hurley, *Billy Mitchell: Crusader for Air Power* (Bloomington: Indiana University Press, 1964), 86–88.
65. Air Corps Board, "Study No. 9—Radio Controlled Aircraft" (1935), 7, as cited in Gillespie *Weapons of Choice*, 22–24.
66. Kenneth Werrell, *Chasing the Silver Bullet: US Air Force Weapons Development from Vietnam to Desert Storm* (Washington, DC: Smithsonian Institute Press, 2003), 139–140.
67. Steven Zaloga, *V-1 Flying Bomb 1942–52: Hitler's Infamous "Doodlebug"* (London: Bloombury Publishing, 2011); Werrell, *The Evolution of the Cruise Missile*, 46.
68. The Navy ordered approximately 150 of the JB-2s, designated the "Loon," which were intended as part of a Japan invasion campaign. Japan surrendered in 1945 and these missiles were never used, but they influenced early Navy cruise missile development post–World War II. Polmar and O'Connell, *Strike from the Sea*.
69. Neufeld, *The Development of Ballistic Missiles in the United States Air Force, 1945–19(?)*; George Mindling and Robert Bolton, *U.S. Air Force Tactical Missiles, 1949–1961* (Morrisville, NC: LuLu.Com, 2008).
70. Polmar and O'Connell, *Strike from the Sea*.
71. Theodore von Karman, *Prophecy Fulfilled: "Toward New Horizons" and Its Legacy Fulfilled* (Maxwell: Air Force History and Museums Program, 1994), 9.
72. Youngblood, *The Development of Mine Warfare*, 110–126.
73. A. G. Solyanking, M. V. Pavlov, and I. V. Pavlov, *Domestic Armored Vehicles, Twentieth Century: 1905–1941* (Moscow: Elksprint, 2002).
74. Everett, *Unmanned Systems of World War I and II*.
75. Ibid., 489.
76. Melvin R. Laird, Secretary of Defense, Testimony before the Committee on Appropriations, House of Representatives, February 25, 1970.
77. Youngblood, *The Development of Mine Warfare*, 143.
78. Scott C. Truver, "Naval Mines and Mining: Innovating in the Face of Benign Neglect," CIMSEC.org, December 20, 2016, https://cimsec.org/naval-mines-mining-innovating-ace-benign-neglect/.
79. G. Harry Stine, *ICBM: The Making of the Weapon That Changed the World* (New York: Orion Books, 1991).
80. Ibid., 148.
81. Max Rosenberg, *The Air Force and the National Guided Missile Program, 1944–1950*, (Newtown: Defense Lion Publications, 2011), 18.
82. Ibid, 36.
83. Stine, *ICBM*, 148.
84. Neufeld. *The Development of Ballistic Missiles in the United States Air Force, 1945–1960*
85. Robert L. Perry, "Appendix," in *Science, Technology, and Warfare: The Proceedings of the 3rd Military History Symposium*, ed. Monte Wright and Lawrence Paszek (Colorado Springs: US Air Force Academy, 1969), 119–121.

86. Robert Art and Stephen Ockenden, "The Domestic Politics of Cruise Missile Development," in *Cruise Missiles Technology, Strategy, Politics*, ed. Richard Betts (Washington, DC: Brookings Institution, 1981), 359–413.
87. Edmund Beard, *Developing the ICBM* (New York: Columbia University Press, 1976); Michael Brown, *Flying Blind: The Politics of the U.S. Strategic Bomber Program* (Ithaca, NY: Cornell Studies in Security Affairs, 1992).
88. Maj. Gen. Thomas Power, Vice Commander, SAC, Letter to Director of Requirements, HQ/USAF, March 30, 1953.
89. Brown, *Flying Blind*, 67.
90. David K. Stumpf, *Minuteman: A Technical History of the Missile That Defined American Nuclear Warfare* (Fayetteville: University of Arkansas Press, 2021); Stine, *ICBM*.
91. Beard, *Developing the ICBM*.
92. Stumpf, *Minuteman*.
93. Morton Halperin, "The Gaither Committee and the Policy Process," *World Politics* 13, no. 3 (1961): 360–384; *Deterrence and Survival in the Nuclear Age (the Gaither Report of 1957)*, Joint Committee and Defense Production Congress of the United States, 94th Congress (Washington, DC: US Government Printing Office, 1976).
94. Stumpf, *Minuteman*.
95. Desmond Ball, *Politics and Force Levels* (Berkeley: University of California Press, 1980); Brown, *Flying Blind*, 241; Stumpf, *Minuteman*.
96. Stumpf, *Minuteman*.
97. Thomas B. Allen and Norman Polmar, *Rickover: Father of the Nuclear Navy* (Washington, DC: Potomac Books, 2007).
98. Stine, *ICBM*, 206–211.
99. Werrell, *The Evolution of the Cruise Missile*, 92. There were also reports of an errant Snark launch that ended up in Brazil.
100. Ibid., 106.
101. Ibid., 126–128.
102. Ibid., 108.
103. Polmar and O'Connell, *Strike from the Sea*.
104. Ibid.; Werrell, *Evolution of the Cruise Missile*, 119.
105. Werrell, *The Evolution of the Cruise Missile*, 150–151.
106. Robert L. Pfaltzgraff Jr. and Jacquelyn K. Davis, *The Cruise Missile: Bargaining Chip or Defense Bargain* (Cambridge: Institute for Foreign Policy Analysis, 1977); Colin Gray, "Who's Afraid of the Cruise Missile?," *Orbis*, Fall 1977, 519; Ronald Huisken, *The Origin of the Strategic Cruise Missile* (New York: Praeger, 1981).
107. "Air Force, Navy to Develop Cruise Missile," *Aviation Week*, August 20, 1973, 24.
108. Werrell, *The Evolution of the Cruise Missile*, 150.
109. Ibid., 135.
110. Brown, *Flying Blind*; Art and Ockenden, "The Domestic Politics of Cruise Missile Development, 1970–1980."
111. Brown, *Flying Blind*.
112. Charles Mohr, "Cruise Missile Passes Test but Its Critics Score, Too," *New York Times*, July 17, 1983; "Pressure for New Bomber Rises in Congress," *Aviation Week and Space Technology*, February 11, 1980.
113. Betts, ed., *Cruise Missiles Technology, Strategy, Politics*; Brown, *Flying Blind*.
114. Youngblood, *The Development of Mine Warfare*, 145–148.
115. Ibid., 150–151.
116. Chrysler Corporation Missile Division, *This Is Redstone* (Periscope Film LLC, 2012).
117. Stine, *ICBM*.
118. John C. Lonnquest and David F. Winkler, *To Defend and Deter: The Legacy of the United States Cold War Missile Program* (Washington, DC: USACERL Special Report 97/01, 1996), 3.
119. Michael James Young, "The US Air Force's Long Range Detection Program and Project MOGUL," *Air Power History* 67, no. 4 (2020): 25–32.
120. Curtis Peebles, *High Frontier: The United States Air Force and the Military Space Program* (Maxwell, AL: Air Force History and Museums Program, 1997).
121. Robert Guerriero, *Space-Based Reconnaissance: From a Strategic Past to a Tactical Future* (Huntsville, AL: Army Space and Missile Defense Command, 2002).

122. Peebles, *High Frontier*, 8.
123. Greg Thielmann, "Looking Back: The Missile Gap Myth and its Progeny," *Arms Control Today* 41, no. 4 (2011): 44.
124. John Deutch in discussion with the authors, April 2020.
125. Thomas Erhard, *Air Force UAVs: The Secret History* (Arlington, VA: Mitchell Institute Press, 2010).
126. Erhard, *Air Force UAVs*; Mark Wells, *Tribal Warfare: The Society of Modern Airmen* (Maxwell Air Force Base, AL: Air University, 2015).
127. Though the Air Force (under McNamara's direction) did also experiment with manned space missions as well as a military space base. These ideas were ultimately rejected. Peebles, *High Frontier*, 15.
128. Though the Air Force would later take over some of the Navy navigation missions after the 1970s and eventually lead the GPS program. Ibid., 32.
129. Ibid., 58.
130. Werrell, *Chasing the Silver Bullet*, 141.
131. Stumpf, *Minuteman*.
132. M. R. Kelley, "The Terrier: A Capsule History of Missile Development," *APL Technical Digest* 4, no. 6 (1965): 18–26.
133. Elliot Converse, *History of Acquisition in the Department of Defense*, vol. 1, *Rearming for the Cold War, 1945–1960* (Washington, DC: Office of the Secretary, Historical Office, 2012).
134. Werrell, *Chasing the Silver Bullet*, 142.
135. Andrew Cockburn, *Kill Chain: The Rise of the High-Tech Assassins* (London: Picador Books, 2015); Werrell, *Chasing the Silver Bullet*, 37.
136. Whitmore, *Evolution of Unmanned Aerial Warfare*.
137. Werrell, *Chasing the Silver Bullet*, 33.
138. Ibid., 144–145.
139. Kenneth Mayer, *The Development of the Advanced Medium-Range Air-to-Air Missile: A Case Study of Risk and Reward in Weapon System Acquisition* (Santa Monica, CA: RAND Corporation, 1994), 15.
140. Gillespie, *Weapons of Choice*.
141. Erhard, *Air Force UAVs*; Cargill R. Hall, "Reconnaissance Drones: Their First Use in the Cold War," *Air Power History* 61, no. 3 (2014): 20–27; Whitmore, *Evolution of Unmanned Aerial Warfare*.
142. Werrell, *Chasing the Silver Bullet*, 33.
143. Polmar and O'Connell, *Strike from the Sea*.
144. "Battle of Palmdale," accessed December 31, 2023, http://415vva.homestead.com/Mil_Hist_Battle_of_Palmdale.pdf.
145. Alice Hunt Friend, "Creating Requirements: Emerging Military Capabilities, Civilian Preferences, and Civil-Military Relations" (PhD diss., American University, 2020); Bill Yenne, *Attack of the Drones: A History of Unmanned Aerial Combat* (Saint Paul, MN: MBI Publishing Company, 2004).
146. Douglas W. Gage, "UGV History 101: A Brief History of Unmanned Ground Vehicle (UGV) Development Efforts," *Unmanned Systems Magazine* 13, no. 3 (1995), 1–10; National Research Council, *Technology Development for Army Unmanned Ground Vehicles* (Washington, DC: National Academies Press, 2003); Glenn Baca, *An Analysis of US Army Unmanned Ground Vehicle Strategy* (Cambridge/Carlisle, PA: MIT/US Army War College, 2012).
147. John Romjue, *A History of Army 86*, vol. 2, *The Development of the Light Division, the Corps, and Echelons above Corps November 1979–December 1980* (Fort Monroe, VA: Army Training and Doctrine Command, 1982).
148. Thomas Mahnken, *Technology and the American Way of War since 1945* (New York: Columbia University Press, 2008), 112; Hall, "Reconnaissance Drones."
149. Richard Whittle, *Predator: The Secret Origins of the Drone Revolution* (New York: Henry Holt and Company, 2014).
150. Ibid.
151. James Roche in discussion with the authors, April 2020.
152. National Defense Authorization, Fiscal Year 2001, H.R. 5308, 106th Congress (2000).
153. *Unmanned Systems Roadmap 2007–2032* (Washington, DC: US Department of Defense, Office of the Secretary of Defense, 2007).

154. National Defense Authorization Act for Fiscal Year 2020, S. 1790, 116th Congress (2019).
155. Ronald O'Rourke, *Navy Large Unmanned Surface and Undersea Vehicle: Background and Issues for Congress*, RL45757 (Washington, DC: US Library of Congress, Congressional Research Service, 2020).
156. Margarita Konaev et al., *U.S. Military Investments in Autonomy and A.I.: A Budgetary Assessment* (Washington, DC: Center for Security and Emerging Technology, 2020), 6.

Chapter 2
Ideas and Unmanned Proliferation

*The world is not the product of technology developing autonomously,
it is not brought about by benignly rational decision making ... [but
by] mundane social processes.*

—*Donald Mackenzie*[1]

Our historical outlay reveals that none of these technologies were inevitable.
Today's contemporary arsenal of drones and remotely piloted aircraft were
not the result of some sort of predetermined linear technological progression, but a labyrinthine pathway of dead ends, rabbit holes, and hidden
passages. This makes the last thirty or so years a particularly interesting puzzle for unmanned proliferation. How did we get here? What drove the rise
of these unmanned technologies in the contemporary US arsenal?

Considering the different paths unmanned technology could have taken,
it is perhaps surprising how little of today's vibrant discussion about
unmanned technology explores *why* unmanned systems proliferate. This is
a shame because the story about the rise of unmanned systems in the US
Department of Defense (DoD) is complicated and messy and fundamentally interesting. Untangling this knot to understand how beliefs translate
into lines within a congressional budget tells us something far beyond a
rote inventory of unmanned systems. It tells us how ideas transit circuitous
paths, how they diverge and intersect based on chance meetings and belief
chaperones, and how some beliefs find themselves in dead ends while others
become ingrained within identities and, ultimately, shape the budgets and
technologies of military arsenals.

You can think of this book, therefore, as a prequel to the work that already
exists on unmanned systems. It explains the beliefs that came a priori to
the capabilities or capacity for contemporary unmanned proliferation. It
develops a cast of characters and tells the back stories—the pivotal moments,
the wars, the budget shocks—that come before structural or capacity explanations for unmanned proliferation. In telling this story, this book also

The Hand Behind Unmanned. Jacquelyn Schneider and Julia Macdonald, Oxford University Press.
© Oxford University Press (2025). DOI: 10.1093/9780190064419.003.0003

creates a polemic about the importance of beliefs to what we might consider the hardest security decisions and argues that decisions to invest in weapons are never made in strictly rational ways. Simply put, it is impossible to explain military power if you divorce the ideational from capacity and structure.

In this chapter, we outline the key beliefs, processes, and pathways that underpin the trajectory of unmanned proliferation in the US DoD. In doing so, we take a broad view of proliferation to include not only the procurement and use of unmanned systems but also their research and development.[2] This allows us to move beyond the conceptualization of proliferation as a dichotomous outcome (do we have unmanned systems or not?) to capture the process of proliferation as a movement from belief to investment and adoption toward an unmanned future across a wide spectrum of possible platforms, munitions, missions, and domains. Most importantly, treating proliferation as a process rather than as a binary allows us to examine trajectories of unmanned development instead of stagnant unmanned orders of battle, and to identify those critical junctures and decision points that created path dependencies for today's unmanned inventory.

In the following section we present the existing rational choice explanation for the proliferation of unmanned technology and explain why it provides an incomplete explanation for the trajectory of unmanned systems. We then turn to academic literature on military innovation and constructivist theories of international relations and foreign policy to explore the role that beliefs play in weapons' proliferation, introducing the concepts of policy entrepreneurs, critical junctures, and path dependencies to show how beliefs infiltrate policy outcomes. With this theoretical backdrop in mind, we then introduce three ideas that we argue shape contemporary US unmanned proliferation—service identity, the belief in technological revolutions, and the belief in casualty aversion—and explore the variety of actors who advocate, adopt, and co-opt these beliefs for their ends. Finally, we lay out our expectations of how we see these ideas influence the trajectory of unmanned systems.

The Rational Choice Perspective—Capacity and Effectiveness

Perhaps the dominant narrative about unmanned proliferation is that states seek to invest in unmanned weaponry when they have the capacity to develop the technology and, secondarily, when their regime type incentivizes risk reduction.[3] These explanations focus on rational, structural incentives for the proliferation of unmanned systems. They take as an assumption that unmanned systems are either roughly similar or more effective than their manned alternatives. Andrea and Mauro Gilli, for example, challenge the view that drone technology will proliferate easily and quickly by focusing on the high technological and infrastructural barriers to states developing unmanned technologies. However, it is assumed that states that have the requisite technological and infrastructural capacity will invest in and field these systems.[4] Matthew Furhmann and Michael Horowitz in their study of drone proliferation similarly find support for capacity-based arguments and highlight technological constraints as limiting the types of unmanned systems states can operate.[5]

If this is true, then beliefs about the use of unmanned systems would be relatively inconsequential variables in the story of US unmanned proliferation, and the trajectory of unmanned systems would be best explained by standard measures of state capacity and development. In particular, we would expect technologies that require less infrastructure and technological know-how to be more widely adopted than technologies that require greater support and knowledge to operate. Following this logic, the explanation for the contemporary adoption of unmanned air systems over, for example, underwater or ground systems is that they require less technological and infrastructural capacity. Compared to unmanned air systems, ground systems operate in highly cluttered environments that create hard problems for autonomy, while underwater systems present an easier problem for navigation but face difficulties communicating through water.

Capacity-based arguments, therefore, seem at first glance to be a good explanation for the current high level of investment in air systems. However, when you expand the lens to look beyond remote-controlled unmanned air systems, these arguments are less compelling. They struggle to account for variation in proliferation: both in similar types across armed services and between different types of unmanned systems even within the same services. Take the unmanned air case. The Navy, for example, was a late adopter of the RQ-4 and has almost no unmanned air strike capability compared to the Air Force. The Army, on the other hand, has focused on small assets, organic to their core warfighting unit—a strong contrast to the strategic and centrally controlled resources preferred by the Air Force. When you expand the question to understand proliferation of systems across mission sets, domains, platforms, and munitions, the capacity argument becomes even less useful. Why, for example, has the Army not invested in more unmanned autonomous vehicles, especially for high-risk convoys or logistical trains in more clutter-free environments like the desert? Why have some precision munitions like long-range air-to-air missiles and air-to-ground bombs proliferated while antiship missiles and ground-based cruise and strategic missiles have almost completely stagnated? Why, despite the ability to automate many of these unmanned air systems, are most of the systems in the US inventory still remote controlled? These questions can't be completely satisfied with a capacity explanation.

Underpinning the state capacity argument is an assumption about military effectiveness and a belief that unmanned technologies will bring revolutionary capabilities to a state's military forces. This kind of argument is based on the presumption that unmanned systems will transform the ways that militaries fight, improving the effectiveness of those that employ these new technologies. In advocating for unmanned technologies, proponents cite a series of characteristics that they believe make unmanned systems uniquely advantaged to future warfare. These include precision, loiter, cost, maneuver, and force protection. Taking their cue from advocates of military revolutions, many scholars that argue for capacity-based explanations for unmanned proliferation believe that these characteristics will exponentially alter states' ability to create and project military power. And because of these assumed benefits, all states will inevitably want to develop and adopt unmanned platforms.

However, it is not clear that any one unmanned system provides the kind of silver-bullet benefits predicted by the revolutionary narrative.[6] In contrast, the extraordinary diversity of characteristics within the family of unmanned systems illustrates how systems designed for one type of battlefield advantage, for example, range, speed, or force protection, often trade off advantages in other battlefield characteristics, like economic cost, control, or resilience. For example, unmanned systems like remotely piloted aircraft that optimize force protection decrease risk to servicemembers but often require a significant logistical tail that is both expensive and cumbersome. Alternatively, unmanned systems—like long-range missiles—designed to create both speed and range create advantages for first movers and quick campaigns. However, the technology to shoot at long ranges, which requires propulsion, fuel, and sophisticated guidance, can be extremely expensive, making it difficult to create mass without going broke. Finally, unmanned systems, like mines, which are both cheap and require very little logistical support, sacrifice control and create dangers for escalation and inadvertent casualties.

The simple truth is that unmanned systems—just like manned systems— have advantages and disadvantages that make them more or less effective at different types of missions, depending on what one seeks to achieve. Which missions are deemed important and the objectives one seeks to achieve are in turn the result of a series of judgment calls that are guided by one's beliefs about politics, the role of armed forces, and how technology can best serve those ends. In other words, to ask whether unmanned or manned systems are more effective misses the point that effectiveness is largely a phenomenon constructed by those who have an interest in using, developing, and sustaining a technology—an idea we will explore more in subsequent chapters.[7]

Beyond the Strictly Rational—The Role of Beliefs in Weapons Proliferation

The problem these strictly rational capacity and structural arguments have in explaining unmanned trajectories is that the choice to invest in unmanned systems is not necessarily about manned versus unmanned. Instead, as our historical outlay demonstrated, there are a wide array of choices that humans may make about how (and why) technology removes humans from the battlefield. So what might explain these choices to invest in different unmanned systems? The military innovation literature offers a series of alternative organizational, cultural, and ideological explanations for when and how military technologies are adopted and used in war. One line of argumentation points to the role of external shocks and key individuals in pushing military organizations to adopt new systems. Barry Posen, for example, emphasizes the rigid nature of military organizations that are purposefully designed to maintain the status quo and avoid dramatic transformation. This conservative bias means that for military organizations to innovate and adopt new technologies, external forces are required to deliver a swift "kick in the pants" to inflexible bureaucrats and military personnel. Defeat in wartime and the presence of interventionist civilian leaders are two such forces that can jolt military organizations and create the conditions necessary for change.[8]

Other scholars have focused on military culture as a key variable in explaining the adoption and use of military technologies. Theo Farrell argues that varying cultural contexts in Europe, America, and East Asia shape military organizations' responses to technological and strategic opportunities, creating different patterns of innovation. These cultural contexts can be shaped and reshaped by senior military personnel by external shocks, or through professional emulation, but once a culture is set, it constrains the choice set and types of responses deemed appropriate by that organization.[9] Similarly, Elizabeth Kier finds that French military doctrine was shaped less by technological trends or geography and more by the combination of civilian elite beliefs about the role of the military in society and military organizational beliefs about its strategic culture. Together, these beliefs about military identity within French society led to the adoption of a defensive doctrine. For Kier, culture shapes the ways in which other variables (like technology) are framed and influence military doctrine. As she writes, "Culture remains relatively static, but constraints set by the independent variables—technology or domestic politics—vary. The organization continues to think along the lines set by its culture and integrates exogenous changes into its established way of doing things. The outcomes (doctrine) may change but the means (its culture) of getting there stay the same."[10] Carvin and Williams argue similarly that culture influences the American way of war. They argue that deep beliefs in liberalism and Enlightenment values, embedded within American culture, explain the American military reliance on science and technology to reconcile a competing desire for "annihilation and restraint."[11]

In addition to state or military culture, the armed services within the military apply their own cultural lenses to innovation. Carl Builder, for example, advances a cultural explanation for military innovation, but one that focuses on the subcultures, or identities, of the different branches of the armed forces. According to Builder, each service has a distinct personality that is shaped by differing histories, traditions, educational pathways, missions, and of course operational environments that lead military personnel from the three services to have varying responses to new technologies. For example, whereas the Navy is imbued with an independent spirit that leads to a preference for independent capabilities, and the Army privileges conventional war fighting over other nontraditional operations,[12] the Air Force has consistently shown a preference for the technological versus human elements of combat.[13]

Others still have disaggregated the military organization further to explore how rank, combat experience, and even the individual perception of their warrior ethos affect proclivities to innovate and adopt new technologies (though the results of this research sometimes suggest contradictory relationships between these individual characteristics and innovation).[14] Barton Hacker, for example, presents evidence from the British interwar years to support the intuition that lower-ranked officers are more receptive to innovation and change than their senior colleagues.[15] Yet more recent research by Thomas Mahnken and James Fitzsimonds suggests that in many instances senior officers are more supportive of new combat methods than their more junior counterparts.[16]

This all leads to a broader argument within a series of case studies on technology adoption in military innovation studies and science and technology studies: that it is the social processes through which beliefs influence technological decisions that explain weapons trajectories. In Donald Mackenzie's explanation of the development of nuclear missile guidance, he argues that the nuclear world "is not the product of technology developing autonomously, it is not brought about by benignly rational decision making" and instead is a social process that shapes nuclear development.[17] Indeed, Graham Spinardi's examination of US ballistic missile development concludes that it was not the technical that drove the weapons, but instead how "technical choices were thus influenced by the 'macropolitics' of US defense policy, by the organizational politics of the Navy and its relationship with the Air Force, and also by the 'micropolitics' of the technical community."[18] Spinardi's work highlights how key individuals and experts interact with these social processes to dictate the bounds of technology proliferation. This is also a recurring theme in explanations of the rise of cybernetic theories, network-centric warfare, and the adoption of digital technologies.[19] As Rebecca Slayton details in her examination of investments in missile defense technology, "Our understanding of the risks and promise of missile defense derives not from self-evident technological realities, not from socially constructed ideas about nuclear war, but from a history of expertise that is simultaneously technological and social."[20]

How do these beliefs interact with processes and expertise to lead to different weapons choices? As Goldstein and Keohane argue, ideas lead to policy outcomes because they "serve as road maps . . . contribute to outcomes in the absence of a unique equilibrium . . . [and] embedded in institutions specify policy in the absence of innovation."[21] Beliefs therefore may play a vital role in shaping the trajectory of emerging technologies like unmanned systems that lack their own road map, institutional equilibrium, or policies. Absent sheets of kill probabilities or damage estimates, defense practitioners in the Pentagon, at combatant commands, and on the battlefield must lean on beliefs to shape the ways in which they experiment, acquire, adopt, and innovate new technologies. Some of these beliefs are about how the weapon will operate, but others are about the nature of war and technology, or the role that identities and institutions play in framing interests and outcomes for defense budgets.

The question remains: where do these beliefs come from? And how do they spread? The answer lies in something akin to the norms "life cycle,"[22] which provides a useful framework for understanding how beliefs can emerge and take hold in a population. The first stage of the life cycle involves norm, or in this case, belief emergence. There are a number of potential sources of beliefs that may affect how they spread through society. Beliefs that arise from personal experience, and especially harrowing experiences such as war, tend to become especially ingrained in individuals, be held more widely, and be more deeply internalized than beliefs that are acquired in less experiential or immersive ways.[23] That is, they assume a taken-for-granted quality that is extremely difficult to replicate.[24] The extraordinarily gruesome effects of chemical weapons in World War I, for example, generated strong beliefs that helped to restrain states during World War II from their use. Nina Tannenwald's work on the nuclear taboo also provides a particularly salient example of how beliefs about the moral repugnance of the use of nuclear weapons led to the development of a norm of nonuse around nuclear weapons.[25]

Beliefs that develop organically from shared experiences may not need individuals to expressly cultivate or champion their ends. Such beliefs tend to permeate collective thinking and shape behavior almost subconsciously and in ways that do not often invite open contestation or debate. In many ways these beliefs quickly assume a norm-like status in the depth of their acceptance—though they may be limited in impact by the small community of shared experience. Over time these "experiential" beliefs tend to seep through society organically, becoming embedded in policies, dominant narratives, and the tacit behavior of generations of individuals— for example, the impact of liberalist ideas in modern Western conflict, an example in which ideas become so ingrained in society that they implicitly but persuasively influence other beliefs about the conduct of war.[26]

By contrast, beliefs that are learned less directly through reading, writing, and observation are less naturally ingrained and may require individual champions acting strategically to promote them.[27] These individuals can act as policy entrepreneurs by investing their "time, energy, reputation, and sometimes money . . . in the hope of a future return" in the form of policies they favor.[28] Their impact is especially pronounced in cases when individuals have the power, resources, and opportunities to push their beliefs forward and generate support from within the policy community. Having the opportunity and resources to advance one's beliefs is often easier the more senior an individual is in a given bureaucracy. This explains the focus in many studies on the role of state leaders' beliefs in policy outcomes.[29]

State leaders are not the only individuals who matter. Other senior decision-makers—secretaries of defense, secretaries of state, national security advisors, leaders of armed services, and more—occupy key positions of power within administrations, have direct access to the president, and can wield substantial influence over policy outcomes.[30] Other individuals are able to inject their beliefs into the policy-making process by gaining the trust and ear of senior decision-makers.[31] Others still become influential due to their expertise on a given subject matter, especially when that subject is particularly complex and requires specialized knowledge to make informed policy decisions.[32] For our story, policy entrepreneurs can exist throughout the DoD, but their job is easier or harder based on their proximity to budget decisions.

Whatever their source of influence, a crucial characteristic of all of these actors is social acuity and the ability to frame beliefs and policy preferences in ways that resonate with the concerns of senior policy makers.[33] Policy entrepreneurs can become more successful in propagating beliefs when they can graft onto existing understandings about war, international relations, or technology.[34] The use of analogies to provide a frame is a common cognitive solution that decision-makers seek when attempting to reason through new situations. Therefore, framing emerging technologies within the lens of familiar previous technologies provides an initial road map for use or implementation.[35] Beliefs that can find common analogies are more likely to take hold for decision-makers and guide their use of emerging technologies. Novel beliefs without an analog will be less likely to be adopted, especially in the initial phase of a technology's emergence.

The presence of enterprising individuals has been used to explain a range of policy outcomes in the United States, from the passing of legislation to reduce greenhouse emissions[36] to the design of child welfare politics.[37] Policy entrepreneurs have also been attributed with bringing about a number of foreign policy outcomes. James Goldgeier highlights the role of key individuals in the Clinton administration in pushing forward North Atlantic Treaty Organization (NATO) enlargement in the 1990s.[38] Charles-Philippe David explores the importance of policy entrepreneurs in transforming US security policy during the G. W. Bush administration.[39] And Mintrom and Luetjens look to policy entrepreneurs to explain the success of the European Union's Common Foreign and Security Policy in the early 2000s as well as the negotiation of the Iran nuclear deal in 2015.[40] In the case of contemporary unmanned technologies, a number of key individuals played an outsized role in advancing beliefs about the acquisition and use of these new systems. In the following pages we will detail the significant impact of Andrew Marshall, Bob Work, Robert Gates, Colin Powell, Ronald Fogleman, and some lesser-known (but no less influential) military personnel, Congress members, and defense wonks on the proliferation of unmanned systems.

The second stage of the life cycle involves a process by which a norm or belief receives broad acceptance among a critical mass of actors, surmounts a tipping point, and then "cascades" through the rest of the population.[41] This can occur through a bottom-up process of gradual infiltration as beliefs become embedded in practices and doctrine, or a top-down process of policy change. However it occurs, it is important to note that belief adoption—whether that be by an individual, firm, organization, or institution—can take place for a multitude of reasons, including true commitment to the belief or its strategic adoption for political, economic, or other ends.[42] Finally, once the belief has become accepted by a sufficient proportion of the population and is held deeply enough, it is internalized and extremely hard to change.[43] At this point the belief is ingrained to a point that it becomes intricately linked to an actor's identity and corresponding accepted (and nonaccepted) behaviors.

Yet despite the demonstrated power of beliefs in shaping decisions, not all beliefs lead to policy change—just like not all norms cascade or emerge. What explains how these beliefs become translated into policy decisions and why those decisions have impact? First, it is important to understand that all beliefs exist and interact with identity, or the "collective aspect of the set of characteristics by which a thing is definitively recognizable or known ... the set of behavioral or personal characteristics by which a person is recognizable as a member of a group."[44] Beliefs can culminate to create or define identities, with some beliefs becoming so important to identities that it can be difficult to explain one without the other. Alternatively, once developed, identities can become more powerful than beliefs—mediating which beliefs are adopted, ignored, or even challenged by individuals.[45] Identity is especially important to explaining the proliferation and success of ideas within the military. From the beginning, incoming recruits are trained to relinquish their pre-military civilian identity and adopt a group identity, as a military member, an airman, a soldier, a sailor, or a marine.[46] This identity can become even more specific as a servicemember spends time within an armed service further defining their military identity within a combat specialty—for example, a fighter pilot, a surface warfare officer, or an infantryman. For new beliefs to influence decisions, they must therefore contend with these identities, with those beliefs that support existing identities most likely to be adopted and those that threaten these identities more likely to be either rejected or ignored.

Beliefs may overcome even identity hurdles when windows of opportunity emerge within what would otherwise be business-as-usual politics to allow for new ideas and beliefs to influence policy streams. These windows of opportunity can open in (at least) two different ways. The first pathway involves windows opening as a result of predictable cycles within the political system, for example, with legislative changes, budget renewals, and rotation in leadership positions and through standard election cycles. Though these are anticipated openings, these events create windows through which ideas and beliefs can influence policy streams and provide opportunities for change.[47]

The second way in which windows can open is from unexpected shocks to the system—unpredictable events that leave politicians reeling and in search of new ideas and policy solutions. These exogenous shocks can be caused by international events such as the fall of the Berlin Wall or the attack on Pearl Harbor, or they can originate domestically with, for example, the assassination of a president, a natural disaster, or a terror attack. Whatever the cause of the shock, these kinds of events often represent critical junctures and forks in the road in decision-making that can lead to various unintended consequences.[48] Policy decisions made at those points in time set in motion processes and patterns of investment that create self-reinforcing processes and seemingly unstoppable path dependencies that have deterministic properties.[49] In Stephen Krasner's words, "Historical developments are path dependent; once certain choices are made, they constrain future possibilities. The range of options available to policymakers at any given point in time is a function of institutional capabilities that were put in place at some earlier period, possibly in response to very different environmental pressures."[50]

In weapons development, this is often due to budget cycles, appropriations, and the financial investment in certain weapons systems at particular moments in time. These decisions may result from a process of inertia, or from the strategic actions of key policy entrepreneurs who seized the opportunity to lock in a particular weapons choice. For example, much of the US focus on offensive weaponry and campaigns in the information revolution can be tied back to path dependencies after a critical juncture in the late 1970s and early '80s in which a major Army doctrinal shift happened in concert with the largest (and last for a decade) investment in defense research and development. This led to investments in "computing [that] enabled Air-Land Battle's offensive doctrines and then locked in the offensive nature of future doctrines because of the dependency on computer-driven offensive weapons. Politics created the technology, but the technology became so powerful that it then necessitated offensive doctrines."[51] The critical juncture of the late 1970s and early '80s set in motion institutional patterns and chains of events that became self-sustaining through budget allocations and investments. Similarly, and as we will see, the budget retrenchment at the end of the Cold War greatly limited the resources available for revolutions in military affairs (RMA) advocates to truly fulfill their vision for unmanned technology.[52]

In sum, there is a large and growing body of research showing that how people and organizations think about technology—that is, the beliefs that they hold about them—are crucial to understanding their use and proliferation. Various actors—policy entrepreneurs, firms, and institutions—cultivate, adopt, and then champion these beliefs in a variety of ways to build support for weapons within the political system. When shocks occur—either from within the system or beyond—windows of opportunity open for these actors to infiltrate the policy space and push forward their ideas, navigating budget cycles and attempting to cement their policy preferences. In doing so, they set in motion path dependencies through which beliefs become translated into decisions about weapons development.

What, then, are the identities and beliefs that have driven unmanned proliferation within the United States? And what do these ideas portend for the trajectory of unmanned systems? We outline the expectations of our story next.

The Ideas Driving Unmanned Proliferation

We argue that two beliefs are crucial to explaining the United States' contemporary unmanned proliferation trajectory and that it is those beliefs' interaction with armed service and occupational identities that ultimately explains when and why some technologies proliferated while others stagnated.

The first belief, military revolutions and technological determinism, holds that technology exists within a linear understanding of history in which technology is an independent agent creating punctuated equilibriums (or revolutions) that increase the effectiveness of militaries that adopt these technologies. Unmanned systems—and now increasingly autonomous systems—are at the apex of this linear historical progression within the information technology military revolution. According to technological determinists, because technology has agency in and of itself, it is the onus of the US military to harness these new and cutting-edge technologies to win eventual competition with its adversaries.

The second set of beliefs is about casualty aversion and force protection and centers around a strong conviction both that the US public is casualty intolerant and that their public support is necessary for strategic military success. According to these beliefs in casualty aversion and force protection, technology substitutes for human risk. In the following section we outline the origins of these beliefs (or the stage of belief emergence) and then discuss how we would expect these beliefs to be promoted and co-opted by key actors within the defense sector—defense firms, Congress members, and especially military services—to advance and impede different unmanned trajectories.

Both these beliefs interact with service and occupational identities that exist a priori to these more contemporary ideas. The ultimate success of these beliefs to not only propagate and cascade but also influence unmanned systems' proliferation is often determined—or is at least mediated by—these beliefs' ability to support these core identities.

Belief Emergence

The Military Revolution and Technological Determinism Belief

One of the primary beliefs that we argue drives US unmanned proliferation is the belief that unmanned weapons are an inevitable next step of technological development. This is intimately tied to a linear understanding of history in which technology brings about revolutions (both societal and in warfare) that leapfrog states above one another in inevitable competition. According to this belief, unmanned technologies are the unavoidable next step for warfare—similar to the adoption of steam engines or gunpowder. States must adopt unmanned systems because those that adopt first are most likely to win wars. The belief comes from a set of analogies linked together to describe historical progress from the development of the longbow to the gun, steam, telegraph, railroad, and mechanization.

Perhaps counterintuitively, the contemporary military revolutions belief within US defense circles originates in the Soviet Union with a series of pieces that began to emerge in the 1970s on the scientific-technological revolution (later adopted by military thought leaders within the Soviet Union as the military-technological revolution). These thought pieces leaned heavily on Marxist beliefs about linear historical progression and the integral role of material variables in society's eventual march toward communism. As the original Soviet authors argued, "In its essence, the scientific-technological revolution expresses the laws of development of the epoch of transition from capitalism to communism, being subjected to them, and is an element of the global process in the sense that it establishes historically the material-technical productions for a communist society."[53] For these Soviet theorists, the potential of automation was a unique advantage of the socialist system as the substitution of material for labor in a socialist society was less likely to lead to upheaval than similar substitutions in capitalist societies.[54] Additionally, Soviet theorists saw these technological upheavals within the frame of revolutions, a concept that fit within Marxist understandings of historical progression and served a domestic party constituency that looked for thought that supported the existence and goals of the Soviet Communist Party.

The idea of a linear progression of history uniquely influenced by technological revolutions spread beyond Soviet political or economic theorists and into Soviet military doctrine.[55] In the mid-to-late 1970s, Soviet Marshal

Nikolei Ogarkov wrote a series of thought pieces introducing the military-technical revolution. The military-technological revolution introduced emerging military technologies to the previous scientific-technological revolution historical framing, arguing that a significant conventional warfare revolution was underway. As early as 1971, Ogarkov notes in *Red Star* that "the fundamentally new types of weapons and combat technology, combined with certain other means, have now become the decisive means for conducting armed combat."[56] A decade later his views about technological revolutions and the historical progression of warfare demonstrated even greater maturity; he writes in 1983 that "A profound and revolutionary—in the full sense of the word—perevorot ['revolution,' 'turn-about,' 'upheaval'] in military affairs is occurring in our time."[57] Later, in a 1984 interview Ogarkov made the direct link between military-technological revolutions and a Marxist perception of history when he quoted Engels's reasoning on military building: "Successes of technology, the moment they have become usable and have been applied in practice in military matters, have immediately—almost forcibly, and often against the will of the military command—caused changes and even revolution in the methods of waging war."[58] Because of these beliefs, Ogarkov championed changes in Soviet doctrine that transitioned away from massive nuclear use to conventional precision strike, made possible only because of emerging capabilities' creation of an "automated reconnaissance."[59]

Ogarkov's ideas didn't stay within the confines of the Soviet Union. Instead, the Marxist understanding of historical progression—and the role that technology played in creating revolutions in societies and militaries—found an ear in Andy Marshall and magnified doctrinal changes already occurring within the United States' Air Land Battle Doctrine. This scientific-technological approach to warfare also fits nicely within modern Western approaches to warfare that linked technology with humanism and liberalism—an implicit belief underlying technological revolutions that technology could make states more likely not only to win wars but also to do so with less violence. Whereas the Soviets linked military-technological revolutions to Marxism, Western adopters could harken to Hegel.[60]

There was therefore already an embedded ideational foundation that made these Soviet ideas not only palatable but also enticing for the American military. This helps explain why a midlevel administrator like Marshall, in charge of an otherwise unremarkably small Office of Net Assessment (ONA) deep in the bowels of the Pentagon bureaucracy, could play an outsized role in idea entrepreneurship within the DoD. In Ogarkov's work, Marshall saw a powerful idea about the future of warfare. He took the ONA's limited budget and invested in thinkers that would over the next thirty years shape not only the trajectory of US defense spending and doctrine but also the narrative of unmanned proliferation. A group of defense wonks would rise in prominence from an early discipleship under Andy Marshall and ONA to the leading defense think tanks in DC and professional military education institutions, eventually rising to lead the DoD. Together this group of ONA civilians pioneered a new strategic narrative for the DoD, the idea of revolutions in military technologies. As Krepinevich wrote in 1994, "Growing evidence exists that over the next several decades, the military systems and operations and, in some respects, the organizations and force structures that dominated the major military establishments during the Cold War will be superseded by new, far more capable means and methods of warfare and by new or greatly modified military organizations. Exploiting this military technical revolution should be an integral part of the Pentagon's long term strategic planning process."[61]

As these individuals rose in the Pentagon, they brought with them the ideas of the original Soviet military-technological revolution, now reframed as RMA. And while RMA peaked (and then temporarily dissipated) under Secretary of Defense Rumsfeld, it re-emerged as a "third offset" of American military technology under Deputy Secretary of Defense Bob Work (a former ONA scholar). These policy entrepreneurs from ONA built new missionary institutions within the DoD that invested heavily in unmanned technology and then shaped a national dialogue on the revolutionary role of unmanned and eventually autonomous weapons. In doing so, they pushed change from inside the Pentagon to influence budgets and acquisitions that would eventually lead to the delegation of unmanned systems to operational units.

Their focus on the linear progression of warfare, punctuated by primarily technological revolutions, also affected which systems they pushed for within the DoD. These thought leaders wanted "revolutionary" effects from unmanned systems, largely tied to a desire to decrease the time required to go from target acquisition to a "kill." Therefore, beliefs in technological determinism led to a preponderance of exquisite unmanned systems that privilege technological innovation over other variables like political risk. In particular, the belief prioritizes unmanned systems that maximize situational awareness and expedite the kill chain to proactively engage with the adversary. For military revolutions beliefs, unmanned systems are not necessarily about creating longer distances between shooter and target, but instead about creating shorter time between American shooters and their targets. They are therefore more likely to be autonomous and less likely to employ remote control. They are also not necessarily platforms, and indeed, the precision-guided munition revolution benefited from the spread of these beliefs.

The Force Protection/Casualty Aversion Belief

But the belief in military revolutions alone cannot explain the current outlay of unmanned systems. A second, often competing, belief also influences the trajectory of unmanned proliferation. This is the belief in public casualty aversion, specifically that public opposition to troop losses means that the US military must prioritize force protection and risk avoidance to achieve campaign success. While casualty aversion beliefs often elide with the narrative of RMA because of technology's ability to create effects while minimizing loss of life, the fundamental motivations behind these beliefs are distinct. For military revolutions, the investment in unmanned technology is about creating revolutionary operational effects. The assumption is that these operational effects will, in the long term, create strategic success. However, with casualty aversion beliefs, strategic success—even with operational success—is impossible without American public support, and casualty aversion beliefs contend that the American public will not support operations unless they minimize casualties. Subsequently, force protection becomes a strategic priority that can even supersede operational success. This leads to investment in unmanned systems that prioritize the mitigation of personnel risk over technologies designed to create an offensive advantage.

We can trace the role of casualty aversion beliefs to Vietnam in which body counts and an increasingly disapproving public decreased the popularity of the Johnson administration and led his successor, Nixon, to adopt operations that mitigated American casualties.[62] An extremely influential study by John Mueller published in 1973 demonstrated the link between media coverage of American losses and the erosion of political support for the campaign.[63] The lesson taken by the US military, especially officers beginning their career as Vietnam was ending, was twofold. First, the lesson from Mueller was that the American public was highly sensitive to casualties and would influence politicians to either curb operations or withdraw from campaigns if casualties became too large or salient. As a consequence, Vietnam-era officers began to view the media as unfriendly amplifiers of casualties; this led subsequent generations of officers to devote significant effort to ameliorating or shaping what would later be known as the "CNN effect" on public support.

The beliefs inculcated within this cadre of officers seemed to be validated with the overwhelming success of the first Gulf War—which featured dramatically low American casualties and extraordinary military success. Subsequently, the Caspar Weinberger doctrine, which debuted in the early 1990s (also colloquially called the Powell doctrine), emphasized the importance of public support to strategic military success and advocated operations that featured political support, low casualties, and overwhelming advantages. The doctrine emerged as the US military was increasingly struggling with its post–Cold War role and debating whether humanitarian intervention was an appropriate (i.e., publicly supported) use of military resources. Especially after the bombing of Marine towers in Lebanon and the very vivid loss of military members in Somalia, force protection became a significant focus in the development of military doctrine and subsequent use of technology in warfare. This collision of casualty aversion, force protection, and technology reached its acme in Bosnia in which the Clinton administration dictated a campaign fought largely with precision-guided munitions and aircraft flying high above anti-aircraft artillery and surface-to-air missile threats. Despite the minimal loss of life in the force-protection-heavy campaign, the shootdown of Scott O'Grady garnered significant media attention and US leaders worried that the potential loss of the fighter pilot could put the entire operation at risk of strategic failure.

Focus on force protection somewhat dissipated at the end of the Bosnian conflict and instead the focus on RMA dominated the doctrinal discussion of US military officers and thought leaders. However, concerns about force protection and casualty aversion weren't rejected but instead internalized inside a more techno-dominant discussion about the future of warfare. Interestingly, it was at this time that academic work on casualty aversion also offered nuance to Mueller's initial finding about public risk aversion to casualties. Studies conducted by Feaver and Gelpi in the mid-1990s as well as a Rand study by Larson in 1996 suggested it wasn't necessarily that the public was always casualty averse, but that their support for military losses coincided with both the context of the situation often provided by elite cues (i.e., was the conflict viewed as strategically important?) and perceptions of military success.[64] Accordingly, the discussion about force protection became more sophisticated and linked technological revolutions with the ability to mitigate risk and ensure public support.

The marriage of technological determinism and casualty aversion helps explain how unmanned aerial vehicles began to dominate US unmanned investments. Not only had the technology matured sufficiently that armed unmanned platforms were a viable possibility by the early 2000s, but also they served the twin goal of protecting US military personnel while pursuing US national security objectives overseas. This meant that political leaders could claim to be meeting the important security needs of the American people while insulating themselves from the domestic political fallout of deploying (or losing) military personnel around the world. And as one might expect, this proved a popular strategy among the US public. Public opinion polls between 2011 and 2015 showed that, on average, the US public maintained a 67% approval rating of the unmanned aerial vehicle program, with only 20% registering disapproval—a consistently high level of support for the use of military force overseas that would have undoubtedly taken a dip if US pilots or ground forces were being placed at risk.[65]

Beliefs in casualty aversion drive a different type of unmanned systems proliferation than military revolutions. In particular, the focus on casualty aversion prioritizes unmanned technology that insulates the most salient forces from risk—fighter pilots, for example. Casualty concerns also lead to unmanned substitutions for similar manned systems as well as platforms that can be used both by an operator and remotely. Finally, a focus on casualty aversion leads to an emphasis on unmanned technologies for low-stakes conflicts, ones in which decision-makers might be concerned that American public opinion would not support the loss of life.

Belief Cascade, Internalization, and Identity—Who Adopts These Beliefs (and Why)?

For beliefs to matter to policy outcomes, they must be adopted by a critical mass of actors to "cascade" through the population and become internalized. There are a wide variety of actors that matter to weapons proliferation— including the military services, civilian defense leaders, Congress, and the defense industrial base. All of these actors interact to ultimately choose what weapons make it into the arsenal and onto the battlefield. However, what our previous chapter illustrates is that the primary pass-through for beliefs, the status quo lens through which weapons decisions are made, is the military services and their respective identities. Beliefs will rarely cascade or internalize—even with a push from the other actors—without buy-in from the military service, a significant external catalyst from the likes of the president, or war.

Let's therefore start first with the role that military services and their identities play in explaining contemporary unmanned investment. Carl Builder's 1989 study on American military strategy, *The Masks of War*,[66] identifies distinct identities within each of the military services. These identities define each of the armed services' attitudes toward technology, based on how technology impacts subservice organizational identities, strategic priorities, and insecurities about service autonomy. Builder, for example, argues that the Navy's focus on tradition and independent command at sea leads to service culture decisions that prioritize the Navy as an institution. As the only service that operates air, sea, and land forces, the Navy has the most subcultures and also views itself as the least reliant service on joint escapades, making it the most traditionally focused of the services. The Air Force, as the newest command, is insecure about its independence and therefore advocates doctrine that emphasizes strategic air power and prioritizes technology over the individual service member. As Builder writes, the Air Force "sees itself as the embodiment of an idea, a concept of warfare, a strategy made possible and sustained by modern technology. The bond is not an institution, but the love of flying machines and flight."[67] In contrast, the Army is focused on personnel and has "roots in the citizenry,"[68] making it a late adopter of technology and an advocate for personnel-heavy doctrine versus the technology-focused efforts of the Air Force.

According to Builder's theory of service identity, the Air Force should be the first to adopt unmanned systems as its identity prioritizes technology over personnel. However, that enthusiasm wanes when unmanned systems threaten the value of (and investments in) manned aircraft and dominant pilot occupational identities. The Air Force identity is therefore constantly grappling with a need to preserve an identity of flight while also securing missions from rival services. This means that the Air Force is willing to adopt and adapt beliefs about technology to fit within its larger identity of flight in order to stave off competition from the Army and the Navy. This identity struggle explains, for example, the Air Force's ultimate willingness to adopt ballistic missiles as an extension of flight, despite initial concern that missiles would threaten manned flight. It also explains why cruise missiles were initially sabotaged and ignored by the Air Force until Congress threatened the service with Navy programs.

In contrast to the Air Force, the Army doesn't have a deep-seated need to claim new technologies or new missions. This is because for the Army, technology is simply an enabler for victory on the ground. There is no inherent need to claim new missions as the focus is more expansively on defending, claiming, and winning territory. As unmanned systems develop and prove their worth in performing the dull and dirty tasks that often impede ground effectiveness, therefore, we would expect Army officers to advocate for unmanned systems—especially for platforms that are not fundamentally tied to their identity. This identity helps explain why the Army may be more likely to adopt unmanned air reconnaissance platforms than unmanned systems that replace ground platforms or manned land missions. Historically, the Army identity has not embraced most unmanned systems—whether they were land platforms or missiles—and even mines (the most prolific unmanned ground system) were embraced only in war and ultimately abandoned as both ineffective and potentially immoral.

Finally, based on Builder's assertions, we should expect the Navy to focus most on unmanned systems that further seapower missions, in contrast to unmanned systems for joint missions. However, not all seapower-focused unmanned systems are likely to receive the same attention. This is because for the Navy, the dominant determinants of weapons adoption are the power of even more specific identities within the Navy—occupational allegiances within surface warfare, aviation, and submarine communities. For unmanned systems to succeed in the Navy, they need a proponent within these occupational identities. This is perhaps best exemplified by the efforts of Admiral Rickover to spearhead nuclear ballistic missiles and later cruise missiles as part of a larger campaign to solidify submarines within Navy budget priorities. The Marines, as a ground subculture under the Navy, are tied to the Navy for budget but more likely to adopt and experiment with unmanned systems in order to compete with the Army for the premier ground warfighting reputation.

Military identity is therefore primarily about service culture but cannot be explained without understanding how occupational identities shift those cultures. For instance, in the Air Force fighter pilots once bandied for influence against bomber pilots, and more recently, pilots of unmanned systems have subdivided into a distinct culture from other manned fighter and bomber platforms. Previous work on operational specialties and support for unmanned systems suggests two hypotheses about how operational specialties may impact the proliferation of unmanned systems. First, specialties that face replacement by unmanned systems may be less likely to adopt them and may actively fight back against the proliferation of these systems through budget choices, doctrine development, and personnel choices. Second, the strength of these occupational specialties in the ultimate trajectories of unmanned systems will be tempered (or magnified) by competitions within the service between occupational identities. When an occupational identity is dominant, and therefore not in competition, it will be more likely to affect unmanned trajectories—either by its support, apathy, or resistance to unmanned systems.

The identity of armed services tends to dominate the trajectory of unmanned systems because the armed services control budget requests and their identities are relatively strong. However, these identities don't operate in an ideational vacuum, nor can they dominate ideas so completely that new ideas don't emerge. We argue that it is how service identities interact with force protection and military revolutions beliefs that explains contemporary investment in unmanned technologies. This is because other actors within the DoD mediate and propagate beliefs that the services then adopt (or challenge). From the combatant commands to the Office of the Secretary of the Defense, think tank experts, and professional military education leaders, all of these civilians impact how the services see themselves—and where they see competitors. Sometimes these civilians are proselytizers, trying to inculcate their preferred beliefs within up-and-coming generations within the services. At other times, civilians dictate beliefs through strategies, policies, and senior officer hiring (and firing). Finally, combatant command may not be directly tasked with equipping the force, but these military leaders adopt technology in combat, devise campaign plans, and sometimes directly engage with technological investment and procurement in overseas contingency funds. The beliefs of combatant commanders about the future of war and military effectiveness therefore mediate service identity—at times conforming with service identities but also directly challenging the technological status quo (especially for those commands that rotate leaders of different services). For example, combatant commanders in long ground campaigns are more likely to advocate force protection beliefs, while combatant commanders in a predominantly naval or air campaign against a near peer adversary may be more likely to push for technologies that conform with military revolutions beliefs.

These challenges to identities by service outsiders are often arbitrated outside of the Pentagon in the halls of Congress. Perhaps the dominant narrative about Congress's role in weapons acquisition is that congressional members have a vested interest in political survival and therefore invest in weapons systems that bring economic revenue to their constituencies.[69] This means that they will want to protect defense industry projects in their constituencies and not advocate for changes to existing programs (unless it is to expand them). This is especially true for very large-platform projects, like ships or aircraft. Munitions (especially those that are cheaper) may be less likely to create large congressional constituencies, and smaller, cheaper sensors or complicated networks may struggle to find support from industries within congressional districts. This is further complicated by supply chains for complicated systems that extend across congressional districts, making multiple districts invested in the success of a system, but without creating so many jobs that they dominate the political priorities of any one congressional member.

But despite the power of this narrative, research suggests that Congress often responds to noneconomic inputs for political survival.[70] Further, despite the predictions of economic constituency arguments, our story of unmanned progression is replete with congressional interventions for munitions, sensors, and platforms that the armed services resisted. Part of this can be explained by the overriding power of ideas to Congress. For example, congressional members respond to voter concerns about the loss of US military personnel in battle, which may temper their support both for budgets that justify bloody wars of attrition and for weapons that seem to propagate those types of campaigns.[71] Further, congressional members may use their limited influence on these wars of presidential choice within legislative power over budgets and oversight—choosing, for example, to cut programs that receive public attention about indiscriminate uses of violence or that needlessly put service members in harm's way.

Finally, congressional members—particularly those that serve on the House and Senate Armed Services Committee—may also act as independent policy entrepreneurs based on their own wartime experiences or prior positions within the Department of Defense. For example, congressional members with experience in Vietnam or other wars of attrition may be more sensitive to the human cost of war (both personally and due to support within their constituencies). This was a large motivator in Senator Warner's push for unmanned systems, while other members' experience within the services made them keenly aware of the effects of service identity on program development. John McCain's experience, for example, enabled him to push back against service requests for status quo weapons like the Air Force's support for the F-35 over other ground-strike aircraft like the A-10, and the Navy's carrier-focused procurement strategies. But Congress can also be more easily persuaded of narratives that are not generated within the armed services as their staffs attend and commission think tank meetings and reports on weapons programs and the future of war. This was a particularly useful means of influence for military revolutions beliefs that were not always adopted by either service leaders or military doctrine developers but were consistently lobbied by think tanks like the Center for Strategic and Budget Analysis (often funded and staffed by ONA acolytes).

The defense industrial base also plays a role in magnifying some of these beliefs—both within Congress and within the services, in order to propagate their own organizational interests. The US defense market is unique insofar as the government represents the entire market for military weapons. There is, in essence, one customer being served by a number of competing firms. The one customer is complex, however, and comprises both the military services that use the weapons and the elected politicians that must pay for them. This makes the budget process, and defense acquisition especially, inherently political. The defense industrial base, once made up of over one hundred firms and now just a handful of major players, develops and assembles the major weapon platforms and supporting technologies that Congress purchases. As private firms, these companies are profit motivated, and that shapes their incentives and investment strategies.[72] While in theory they have to listen to their main customer—the military services—and be responsive to their needs, defense firms' preference will always be to maximize profits, which we would expect to incline them to upgrade existing platforms rather than invest in something new and risky. In the absence of a clear need for a new weapons system—generated, for example, by an imminent security threat—business-as-usual politics should tend toward the status quo.[73] Of the two beliefs, we would expect this to lead firms to co-opt the force protection narrative, privileging incremental improvements of existing unmanned platforms over costly (and risky) investments in new systems. It also leads the defense industrial base to focus on platforms over munitions, especially those that have sticky investments in factories and assembly lines.

So how do these beliefs and different actors come together to explain the trajectory of unmanned systems? How might we expect the following chapters to unfold? We have already outlined how the two different beliefs—in military revolutions and force protection—lead to preferences for investments in different types of unmanned systems. Military revolutions advocates will seek unmanned systems that decrease time between sensor and shooter, that have high-tech-advanced sensors, processors, and communication devices that can augment highly sophisticated networked warfare. Force protection advocates, on the other hand, will prefer to invest in technologies that mimic manned systems for standard mission sets and will opt for incremental upgrades to existing, proven platforms that mitigate risk to human life. These are the remotely piloted, large airframes that dominated the battlefield in the 1990s and early 2000s. These beliefs are promoted, adopted, and internalized by different actors in the defense sector—the military, defense firms, and Congress—each of which holds particular views about the role of unmanned technology in its overarching mission and purpose that both propels forward and stymies the development of unmanned systems at different points in time.

The struggle between these two beliefs should become most apparent (and most consequential) when exogenous shocks to the system create windows of opportunity for new ideas to enter the policy space and opportunities open for individuals to push forward their vision for unmanned systems. Vietnam, the end of the Cold War, the Gulf War, 9/11—all of these events created moments of great uncertainty but also opportunity for policy entrepreneurs like Andy Marshall, Robert Work, or Colin Powell to enter the fray. Even turnovers in administration, while not unpredictable junctures per se, still create moments for change when the arrival of new leadership offers the potential for leaders' beliefs in unmanned technologies to converge on a clear policy outcome.

The struggle that occurs between these two beliefs should not be an entirely even one, however. The origin of the force protection belief in experiences in Vietnam gives it a taken-for-granted quality that makes it a potent force in defense circles, and one that should take on more salience than a carefully propagated technocratic narrative about a future revolution in military affairs. Moreover, the force protection narrative serves the interests of Congress and the defense industrial base, further magnifying its effect. Where these beliefs come into conflict, therefore, we would expect the force protection belief and accompanying unmanned investments to be dominant. That is not to say the military revolutions advocates will always lose—sometimes shocks can be big enough and stars can align. The battle should be an uphill one, though, and we would expect these advocates to lose out more often than not. When this occurs, and one belief loses out to another in the bid for influence, we expect to see policy entrepreneurs continuing to operate in the background, building networks and influence campaigns, co-opting beliefs and narratives when it is strategically useful, and getting ready to pounce when another window of opportunity opens.

Some may ask why we focus on only two core beliefs to explain contemporary unmanned investment. There is, for example, existing significant work on how the ideas of international law, ethics and norms, and liberalism shape US defense decision-making.[74] Our argument does not suggest that these ideas never matter or even that they compete with each other. Instead, this book is a refinement of broader arguments about liberalism, ethics, and the American way of warfare, demonstrating how these larger ideas influence specific contemporary beliefs. Liberalism, for example, is foundational to both technological determinism and casualty aversion and explains why these beliefs can often elide as they seek to solve a fundamental paradox of liberalism and warfare: the desire to both win wars and do so with limited loss of life. From this perspective, it is clear how contemporary unmanned systems—both remote and precise—can trace their dominance to the power of liberalist ideas.

International law and ethics, in contrast, often have a more instrumental relationship with beliefs about unmanned systems, cited when it benefits users and discarded when it doesn't. Historical narratives against unmanned systems often accuse them of being weapons of cowardice, without valor or ethics. However, these arguments are dropped when necessity dictates. For example, the Union's General Sherman was disdainful of the Confederacy's use of land and naval mines until begrudgingly he saw their utility for defending against raiding forces. Indeed, a historical US repugnance for land and naval mines is perhaps better explained by the systems' lack of usefulness for US geopolitics than a deep-seated concern about law or ethics.

Conclusion

The challenge in tracing beliefs to actions is demonstrating how ideas translate into policy or defense decisions. The core beliefs that we introduce in this book are not entirely independent. Indeed, beliefs about force protection often elide into military revolutions that then are more likely to be adopted by some service and occupational specialties than others. Perhaps most importantly for this book, we do not seek to place these beliefs in competition against each other or to show that one belief dominates others in the trajectory toward unmanned platforms. Instead, we argue that existing explanations of unmanned proliferation that focus on structural or material explanations are looking at causal mechanisms that come after, not before, beliefs. The beliefs that we identify in the book drive the assumptions behind rationalist arguments and often come a priori to the capacities that rationalist arguments focus on as causal mechanisms.

In the end, this book is not about identifying a single explanation for the rise of unmanned technology. Instead, this is a book about context and complexity, about how independent and sometimes endogenous variables interact to drive unmanned weapons development. It tells the story of how ideas about warfare, which would at first seem disconnected to the contemporary use of unmanned technologies, became so internalized by individuals and then military institutions that they transition from beliefs to identities. It is a story about how ideas compete with one another and find themselves magnified—and nullified—by budget cycles, industry pressures, and institutional desires. It is a story of how stories and their tellers matter, how narratives solidify beliefs and drive the very capacities that rationalist scholars look to as predictors of unmanned development.

Notes

1. Mackenzie, *Inventing Accuracy*, 4.
2. This is in line with many studies in the nuclear politics literature, which include a number of different stages of development along the proliferation road. See e.g. Sonali Singh and Christopher R. Way, "The Correlates of Nuclear Proliferation: A Quantitative Test," *Journal of Conflict Resolution* 48, no. 6 (December 2004): 859–885; Dong-Joon Jo and Erik Gartzke, "Determinants of Nuclear Weapons Proliferation," *Journal of Conflict Resolution* 51, no. 1 (2007): 167–194; Todd C. Robinson, "What Do We Mean by Nuclear Proliferation," *Nonproliferation Review* 22, no. 1 (2015): 53–70.
3. Fuhrmann and Horowitz, "Droning On"; Gilli and Gilli, "The Diffusion of Drone Warfare?"
4. Gilli and Gilli, "The Diffusion of Drone Warfare?"
5. Fuhrmann and Horowitz, "Droning On."
6. Jacquelyn Schneider and Julia Macdonald, "Looking Back to Look Forward: Autonomous Systems, Military Revolutions, and the Importance of Cost," *Journal of Strategic Studies* 47, no. 2 (2023): 162–184, https://www.doi.org/10.1080/01402390.2022.2164570; Calcara et al., "Will the Drone Always Get Through?"
7. For a broader discussion of this point see Ford, *Weapon of Choice*.
8. Posen, *The Sources of Military Doctrine*; Rosen, *Winning the Next War*.

9. Theo G. Farrell and T. Terriff, *The Sources of Military Change: Culture, Politics, Technology* (Boulder, CO: Lynne Rienner, 2002); Farrell, "Figuring Out Fighting Organizations: The New Organizational Analysis in Strategic Studies," *Journal of Strategic Studies* 19, no. 1 (Spring 1996): 122–135; Farrell, "Culture and Military Power," *Review of International Studies* 24, no. 3 (Fall 1998): 407–416; Peter Mansoor and Williamson Murray, eds., *The Culture of Military Organizations* (Cambridge: Cambridge University Press, 2019).

10. Elizabeth Kier, "Culture and Military Doctrine," *International Security* 19, no. 4 (Spring 1995): 65–93.

11. Stephanie Carvin and Michael John Williams, *Law, Science, Liberalism and the American Way of Warfare* (Cambridge: Cambridge University Press, 2015), 12.

12. See also Eric R. Giordano, *The US Army and Non-Traditional Missions: Explaining Divergence in Doctrine and Practice in the Post-Cold War Era* (PhD diss., Tufts University, 2003).

13. See e.g. Builder, *The Masks of War*, 3, 19–20. For a history of Air Force technological focus, see Neufeld et al., *Technology and the Air Force: A Retrospective Assessment.*

14. Christopher Coker, *The Warrior Ethos: Military Culture and the War on Terror* (New York: Routledge, 2007).

15. See e.g. Barton C. Hacker, "The Military and the Machines: An Analysis of the Controversy over Mechanization in the British Army, 1919–1939" (PhD diss., University of Chicago, 1969) 254–256.

16. Thomas Mahnken and James R. Fitzsimonds, "Revolutionary Ambivalence: Understanding Officer Attitudes toward Transformation," *International Security* 28, no. 2 (2003): 115. This finding also accords with Rosen's argument that military innovation stems from forward-thinking senior military officers. Rosen, *Winning the Next War.*

17. Mackenzie, *Inventing Accuracy*, 4.

18. Spinardi, *From Polaris to Trident*, 168.

19. Antoine Bousquet, *The Scientific Way of Warfare: Order and Chaos on the Battlefields of Modernity* (New York: Columbia University Press, 2009); Thomas Rid, *Rise of the Machines: A Cybernetic History* (New York: W. W. Norton & Company, 2016).

20. Slayton, *Arguments That Count*, 224.

21. Goldstein and Keohane, *Ideas & Foreign Policy*, 12–13.

22. The norm life cycle is explained in detail in Finnemore and Sikkink, "International Norm Dynamics and Political Change."

23. Robert Jervis, *Perception and Misperception* (Princeton, NJ: Princeton University Press, 1976).

24. Peter J. Katzenstein and Nobuo Okawara, "Japan's National Security: Structures, Norms, and Policies," *International Security* 17, no. 4 (1993): 84–118; Judith Goldstein, *Ideas, Interests, and American Trade Policy* (Ithaca, NY: Cornell University Press, 1993); Martha Finnemore, "Norms, Culture, and World Politics: Insights from Sociology's Institutionalism," *International Organization* 50, no. 2 (1996): 325–334; Audie Klotz, *Norms in International Relations: The Struggle against Apartheid* (Ithaca, NY: Cornell University Press, 1999); Peter J. Katzenstein, *The Culture of National Security: Norms and Identity in World Politics* (New York: Columbia University Press, 1996); Thomas Risse-Kappen, Stephen C. Ropp, and Kathryn Sikkink, eds., *The Power of Human Rights: International Norms and Domestic Change*, vol. 66 (New York: Cambridge University Press, 1999); Michelle Jurkovich, "What Isn't a Norm? Redefining the Conceptual Boundaries of 'Norms' in the Human Rights Literature," *International Studies Review* 22, no. 3 (2020): 693–711; John Mueller, "The Impact of Ideas on Grand Strategy," in *The Domestic Bases of Grand Strategy*, ed. Richard N. Rosencrance and Arthur A. Stein (Ithaca, NY: Cornell University Press, 1993), 48–62.

25. Tannenwald, *The Nuclear Taboo*. See also John L. Campbell, "Ideas, Politics, and Public Policy," *Annual Review of Sociology* 28 (2002): 21–38. For instance, authors have identified how normative perceptions of policies on affirmative action can create consensus about legislation that doesn't benefit particular legislatures. See e.g. John David Skrentny, *The Ironies of Affirmative Action: Politics, Culture, and Justice in America* (Chicago: University of Chicago Press, 2018).

26. Christopher Coker, *Humane Warfare* (New York: Routledge, 2001); Carvin and Williams, *Law, Science, Liberalism, and the American Way of Warfare.*

27. In fact, the military innovation literature has long noted the important role of individuals in explaining variation in military change. Stephen Rosen, for example, attributes instances of military innovation to the actions of entrepreneurial senior officers who advocate for the

adoption of new doctrinal concepts. Barry Posen also highlights the role of senior civilian leaders who, in conjunction with "maverick" military officers, are able to push forward new, innovative ideas within otherwise conservative bureaucracies. More recently, Adam Jungdahl and Julia Macdonald have noted that civilian and military leaders can act both as agents of change and forces of resistance within military bureaucracies. The failure of the Union Army to acquire breech-loading and repeating rifles en masse during the American Civil War, for example, is explained in large part by the chief of ordnance General Ripley's belief that this technological development encouraged poor discipline among US soldiers. See Posen, *Sources of Military Doctrine*; Rosen, *Winning the Next War*; Jungdahl and Macdonald, "Innovation Inhibitors in War." The increasing interest in individuals within the military innovation literature has been accompanied by a broader behavioral revolution within the social sciences. See e.g. Emilie M. Hafner-Burton et al., "The Behavioral Revolution and International Relations," *International Organization* 71 (Supplement 2017): S1–S31. The entire issue was devoted to this topic.

28. Kingdon, *Agendas, Alternatives, and Public Policies*; Mintrom, *Policy Entrepreneurs and School Choice*; Mintrom and Norman, "Policy Entrepreneurship and Policy Change," *Policy Studies Journal* 37, no. 4 (2009): 649–667.

29. See e.g. Elizabeth Saunders, *Leaders at War: How Presidents Shape Military Interventions* (Ithaca, NY: Cornell University Press, 2011); Rachel Whitlark, *All Options on the Table* (PhD diss., George Washington University, 2014).

30. Jean Garrison, *Games Advisors Play: Foreign Policy in the Nixon and Carter Administrations* (College Station: Texas A&M University Press, 1999).

31. David J. Rothkopf, *Running the World: The Inside Story of the National Security Council and the Architects of American Power* (New York: PublicAffairs, 2005); Joseph G. Bock, *The White House Staff and the National Security Assistant: Friendship and Friction at the Water's Edge*, Contributions in Political Science, no. 170 (Westport, CT: Greenwood Press, 1987).

32. See e.g. Julia M. Macdonald, "Eisenhower's Scientists: Policy Entrepreneurs and the Test Ban Debate 1954–1958," *Foreign Policy Analysis* 11, no. 1 (2015): 1–21; Elizabeth N. Saunders, "No Substitute for Experience: Presidents, Advisers, and Information in Group Decision Making," *International Organization* 71 (Supplement 2017): S219–S247; Emmanual Adler, "The Emergence of Cooperation: National Epistemic Communities and the International Evolution of the Idea of Nuclear Arms Control," *International Organization* 46, no. 1 (Winter 1992): 101–146. On the role that expertise plays in successful persuasion see Carl Hovland, Irving L. Janis, and Harold H. Kelley, *Communication and Persuasion* (Westport, CT: Greenwood Press Publishers, 1953).

33. Mintrom and Norman, "Policy Entrepreneurship and Policy Change," 652.

34. Martha Finnemore and Kathryn Sikkink, "International Norm Dynamics and Political Change"; Goddard, "Brokering Change"; Acharya, "How Ideas Spread"; Finnemore and Hollis, "Constructing Norms for Global Cybersecurity."

35. Robert Jervis, *Perception and Misperception in International Politics*; Richard Ned Lebow, "Miscalculation in the South Atlantic: The Origins of the Falkland War," *Journal of Strategic Studies* 6, no. 1 (1983): 5–35; Stephen Benedict Dyson and Thomas Preston, "Individual Characteristics of Political Leaders and the Use of Analogy in Foreign Policy Decision Making," *Political Psychology* 27, no. 2 (2006): 265–288; David Patrick Houghton, "Historical Analogies and the Cognitive Dimension of Domestic Policymaking," *Political Psychology* 19, no. 2 (1998): 279–303.

36. Barry Rabe, *Statehouse and Greenhouse: The Stealth Politics of America Climate Change Policy* (Washington DC: Brookings Institution Press, 2004).

37. Jacelyn Elise Crowley, *The Politics of Child Support in America* (New York: Cambridge University Press, 2003).

38. James Goldgeier, *Not Whether but When: The US Decision to Enlarge NATO* (Washington DC: Brookings Institute Press, 1999).

39. Charles-Philippe David, "Policy Entrepreneurs and the Reorientation of National Security Policy under the G.W. Bush Administration (2001–04)," *Politics & Policy* 43, no. 1 (2015): 163–195.

40. Michael Mintrom and J. Leutjens, "Policy Entrepreneurs and Foreign Policy Decision Making," in *Oxford Research Encyclopedia of Politics*, May 24, 2017, accessed December 31, 2023, https://doi.org/10.1093/acrefore/9,780,190,228,637.013.463.

41. Cass Sustein, *Free Markets and Social Justice* (New York: Oxford University Press, 1997).
42. On this point see Frank Schimmelfennig, "The Community Trap: Liberal Norms, Rhetorical Action, and the Eastern Enlargement of the European Union," *International Organization* 55, no. 1 (2001): 47–80.
43. Finnemore and Sikkink, "International Norm Dynamics and Political Change."
44. Rogers Smith, "Identities, Interests, and the Future of Political Science," *Perspectives on Politics* 2, no. 2 (2004), 301–312.
45. William O. Chittick, Keith R. Billingsley, and Rick Travis, "A Three-Dimensional Model of American Foreign Policy Beliefs," *International Studies Quarterly* 39, no. 3 (1995): 313–331; James D. Fearon, "What Is Identity (as We Now Use the Word)" (unpublished manuscript, Stanford University, 1999); Katzenstein, *The Culture of National Security.*
46. Rino Bandlitz Johansen, Jon Christian Laberg, and Monica Martinussen, "Military Identity as Predictor of Perceived Military Competence and Skills," *Armed Forces & Society* 40, no. 3 (2014): 521–543; Steven L. Lancaster and Roland P. Hart, "Military Identity and Psychological Functioning: A Pilot Study," *Military Behavioral Health* 3, no. 1 (2015): 83–87; Volker Franke, *Preparing for Peace: Military Identity, Value Orientations, and Professional Military Education* (New York: Greenwood Publishing Group, 1999).
47. Kingdon, *Agendas, Alternatives, and Public Policies*, 213.
48. Critical junctures are, by definition, contingent events that are unpredictable or random. James Mahoney and Celso M. Villegas, "Historical Enquiry and Comparative Politics," in *The Oxford Handbook of Comparative Politics*, ed. Carles Boix and Susan Stokes (New York: Oxford University Press, 2007), 79–80.
49. Ira Katznelson, "Periodization and Preferences," in *Comparative Historical Analysis in the Social Sciences*, ed. James Mahoney and Dietrich Rueschemeyer (Cambridge: Cambridge University Press, 2004), 290; James Mahoney, "Path Dependence in Historical Sociology," *Theory and Society* 29, no. 4 (2000): 507–548.
50. Stephen Krasner, "Sovereignty: An Institutional Perspective," *Comparative Political Studies* 21, no. 1 (1988): 67.
51. Jacquelyn Schneider, *The Information Revolution and International Stability: A Multi-Article Exploration of Computing, Cyber, and Incentives for Conflict* (PhD diss., George Washington University, 2017).
52. On the politics of defense spending, see Adams and Williams, *Buying National Security.*
53. Chelovek-nauka-tekhnika (Moscow: Politizdat, 1973), 352, unattributed English translation, *Man, Science, Technology: A Marxist Analysis of the Scientific and Technological Revolution* (Moscow-Prague: Academia, 1973), 387, as cited in Erik P. Hoffman, "Review: Soviet Views of 'The Scientific-Technological Revolution,'" *World Politics* 30, no. 4 (1978), 615–644.
54. "The System of Power and Democratic Institutions," *Social Sciences (Moscow)* 6, no. 3 (1975) 122; G. N. Volkov, *Man and the Challenge of Technology* (Moscow: Novosti, 1972), 2.
55. Dima P. Adamsky, "Through the Looking Glass: The Soviet Military Technical Revolution and the American Revolution in Military Affairs," *Journal of Strategic Studies* 31, no. 2 (2008): 257–294; Timothy Waddell, *Marshal N.V. Ogarkov and the Transformation in Military Soviet Affairs* (thesis, University of Manitoba, 1999), https://mspace.lib.umanitoba.ca/ bitstream/handle/1993/19,458/Waddell_Marshal_N.pdf? sequence=1; Mary C. Fitzgerald, *Marshal Ogarkov on Modern War: 1977 to 1985* (Washington DC: Center for Naval Analyses, 1987).
56. MSU N. V. Ogarkov, "A Reliable Defense to Peace," *Krasnaya zvezda* (hereafter cited as *KZ*), September 23, 1983; as cited in Fitzgerald, *Marshal Ogarkov on Modern War*, 8.
57. As cited in Fitzgerald, *Marshal Ogarkov on Modern War*, 55.
58. CIA, "USSR National Affairs Political and Social Developments," May 9, 1984, https://www. cia.gov/library/readingroom/docs/CIA-RDP90T00155R000500030019-6.pdf.
59. Ibid.
60. Carvin and Williams, *Law, Science, Liberalism and the American Way of Warfare*; Coker, *Waging War Without Warriors?*
61. Andrew F. Krepinevich Jr., "Keeping Pace with the Military-Technological Revolution," *Issues in Science and Technology* 10, no. 4 (1994): 23–29.
62. John A. Gentry, "Casualty Management: Shaping Civil-Military Operational Environments," *Comparative Strategy* 30, no. 3 (2011): 242–253, https://www.doi.org/10.1080/01495933. 2011.561737.

63. John E. Mueller, *War, Presidents, and Public Opinion* (New York: Wiley, 1973).
64. Peter Feaver and Christopher Gelpi, *Choosing Your Battles: American Civil-Military Relations and the Use of Force* (Princeton, NJ: Princeton University Press, 2005); Eric Victor Larson, *Casualties and Consensus: The Historical Role of Casualties in Domestic Support for US Military Operations* (Santa Monica, CA: Rand Corporation, 1996).
65. Grant M. Cohen, *Origins of U.S. Public Opinion for Drone Strikes: The Intersection of Elite Rhetoric, Media Coverage, and American Public Opinion, 2000–2015* (PhD diss., University of Miami, 2018), 87.
66. Builder, *The Masks of War*.
67. Ibid., 32.
68. Ibid., 19.
69. Larry M. Bartels, "Constituency Opinion and Congressional Policy Making: The Reagan Defense Buildup," *American Political Science Review* 85, no. 2 (1991): 457–474; Ralph G. Carter, "Senate Defense Budgeting, 1981–1988: The Impacts of Ideology, Party, and Constituency Benefit on the Decision to Support the President," *American Politics Quarterly* 17, no. 3 (1989): 332–347.
70. James Lindsay, "Parochialism, Policy, and Constituency Constraints: Congressional Voting on Strategic Weapons Systems," *American Journal of Political Science* 24, no. 4 (1990): 936–960; Barry Rundquist and Thomas M. Carsey, *Congress and Defense Spending: The Distributive Politics of Military Procurement*, vol. 3 (Norman: University of Oklahoma Press, 2002).
71. Gholz and Sapolsky, "Congress and the Defense Industrial Base"; Gholz, Sapolsky, and Talmadge, *US Defense Politics*, 65.
72. Gholz and Sapolsky, "Congress and the Defense Industrial Base," 2.
73. Ibid., 5. See also Christian Brose, *The Kill Chain: Defending America in the Future of High-Tech Warfare* (London: Hachette UK, 2020).
74. Carven and Williams, *Law Science, Liberalism and the American Way of Warfare*; Coker, *Waging War without Warriors*; Colin Kahl, "In the Crossfire or the Crosshairs? Norms, Civilian Casualties, and US Conduct in Iraq," *International Security* 32, no. 1 (2007): 7–46; Kahl, "Constructing a Separate Peace: Constructivism, Collective Liberal Identity, and Democratic Peace," *Security Studies* 8, no. 2–3 (1998): 94–144; Michael Glennon, "Pre-empting Proliferation: International Law, Morality, and Nuclear Weapons," *European Journal of International Law* 24, no. 1 (2013): 109–127.

Chapter 3
Military Revolutions and Technological Determinism

The future of war will be robotic.

—*Peter Singer*[1]

In 1971, sitting in an unremarkable office with no authority to acquire or develop technology, create campaign plans, or generate strategy, a civilian in charge of the small Office of Net Assessment (ONA) commissioned a study on an emerging Soviet concept of scientific-technological revolution. And so began the genesis of one of the most influential beliefs to drive the trajectory of unmanned development in US defense: the revolution in military affairs (RMA) and the inevitability of technological progression.

The narrative of inevitability is interwoven into the lexicon around unmanned systems. The former director of tactical technology at the Defense Advanced Research Projects Agency asserted that unmanned technologies have no choice but to proliferate because unmanned "is the natural evolutionary path."[2] Drone technical conventions lure defense buyers in with promises of progress, telling them, "'Welcome to the future' [in which the drone] is the 'aerial apex,' which will 'become an invaluable tool,' the 'clear' or 'obvious' choice for a plethora of solutions, and "the natural evolution of aerial technology.'"[3]

Where did these beliefs about the inevitability of unmanned systems come from? Here's where our story returns to the bowels of the Pentagon because these contemporary discussions are tied intimately to a perception of history that goes back to the Soviet materialist philosophies first discovered by Andy Marshall in ONA. According to this perspective, technology determines change in societies, economies, and, most importantly for the Department of Defense (DoD), warfare. For those who promote unmanned systems as revolutionary progressions in modern warfare, it is therefore important to understand history and warfare as a linear progression in which technology

The Hand Behind Unmanned. Jacquelyn Schneider and Julia Macdonald, Oxford University Press.
© Oxford University Press (2025). DOI: 10.1093/9780190064419.003.0004

creates an evolutionary wave crashing toward the future.[4] For these schol-ars and practitioners, unmanned technologies *must be* developed because they are the next step in the forward progress of warfare—as unavoidable and disruptive as the development of gunpowder, the steam engine, or mechanization.

How did this belief in military revolutions and technological determinism develop? How does it affect the trajectory of unmanned systems? Who were the key players and institutions, and what were the path dependencies that made the belief in technological progress, RMA, and unmanned inevitabil-ity so important to DoD investment decisions? To answer these questions, this chapter will first introduce the military revolution narrative and trace the development of the RMA belief as well as its underlying relationship with technological determinism. We will then explore how unmanned sys-tems fit within the RMA narrative. Finally, we will examine how the belief emerged, cascaded, and internalized to ultimately influence the trajectory of unmanned systems in today's US military.

The Historical Narrative of the Military Revolution

The belief in military revolutions can be traced to understandings of war-fare and history that emerged in US defense circles at the end of the Cold War.[5] These beliefs reached their apex in the early 2000s as the RMA, later transforming into the third offset strategy and undergirding great power competition.[6] According to this understanding of history and pro-gression of warfare, "Military revolutions recast society and the state as well as military organizations. They alter the capacity of states to create and project military power. And their effects are additive. States that have missed the early military revolutions cannot easily leapfrog to success in war by adopting the trappings of military technology."[7] History, therefore, is a progression of military revolutions in which technologies, in conjunction with societal, doctrinal, and tactical adaptations, make states qualitatively better at warfare than their adversaries. Much like Darwin's evolution-ary theory, the states that adapt best to these revolutions survive, while those that fail to adapt become either obsolete or irrelevant to international politics.

Technology plays a pivotal role in this evolutionary history narrative. The arrival of revolutionary technologies creates an impetus for a punctuated equilibrium within history. Prior to the arrival of the revolution, states exist in a stasis in which there are relatively few comparative advantages. The state that is able to devote the most resources to warfare is also the most likely to win. However, the RMA literature suggests that some technologies allow states that are first movers to disrupt the equilibrium and surpass adversaries—even independent of their overall investment in defense. From the longbow to gunpowder, steam, the telegraph, the railroad, and the tank, states able to harness the revolutionary technology—both societally and militarily—will vault to the top of the international system: "Technology sets the parameters for the possible, it creates the potential for a military revolution."[8]

The corollary, of course, is that those who fail to adapt quickly enough will decrease in power. The technology and warfare revolution literature highlights these evolutionary missteps in warfare: "History is full of examples of superpowers failing to take advantage of important Revolutions in Military Affairs: the Mongols missed the Gunpowder Revolution; the Chinese, Turks, and Indians missed the Industrial Revolution; the French and British missed major parts of the Second Industrial Revolution; the Soviets missed the Information Revolution."[9] Indeed, the idea of the bumbling officer, unable to see the impact of new technology on warfare, is a constant theme in the military revolution literature. No one wanted to be Field Marshal Haig, who has been quoted (perhaps errantly) in a series of military revolutions analyses for a 1925 unfortunate assessment: "You find just as much use for the horse—the well-bred horse—as you have ever done in the past."[10]

Underlying much of this literature on military revolutions is a linear, evolutionary understanding of the progression of warfare. For example, the Tofflers' futurist work that circulated in the DoD during the 1980s and '90s argued that there were social waves "propelled by new kinds of productive activity, which function as the latest and most potent sources of wealth creation and allegedly find unique military expression in characteristic styles of warfare."[11] The future of war in these accounts was an evolutionary progression driven by technology—separate from human agency in which technology exists as a unique and immutable force (i.e., an invisible hand marching progress forward). As Van Creveld explains, "Technology, then . . . might be described as linear. . . . [B]eginning in prehistoric times, it was this uniform, repetitive, predictable character of physical nature which made technology possible. The only reason why even a simple technology, a hammer for example, can be constructed is because we are certain that its effect on a nail will always be the same."[12]

While the military revolutions narrative suggests a linear progression of history and leans toward technological determinism, it should be noted that the question about how much agency technology has over revolutionary progress is an open question within the military revolution literature.[13] Jeremy Shapiro's late 1990s analysis of information and the RMA suggested that the historical approach toward the understanding of warfare reduced the role of human decision-making in its development and, perhaps subconsciously, led to a belief within much of the military revolution literature that warfare progressed independent of human choices. For many, though the technology was vitally important to the progression of warfare, it was incomplete without nontechnological adaptation (or transformation). As Murray explains, "One persistent assumption about military technology (sometimes even championed by technologists) is that it creates requirements on its own; innovation, therefore, occurs when the tools of war demand that military organizations adopt and use them. This 'technological imperative' argument often includes a coda: if we do not exploit this weapon, others will, so at least we must develop it in order to study it. Certainly technological possibilities shaped interwar innovation, but never provided dei ex machina. Instead, the third consideration, the political behavior of military organizations, played a pivotal role in encouraging, retarding, and channeling technology-based innovations."[14] Murray's perspective was pivotal to what would become the dominant American narrative about military revolutions (spearheaded by ONA). Technology would likely provide the catalyst for a move up the linear progression of warfare, but it would be the states that adapted their organizations and societies to that innovation that would "win" the revolution. This would lead to one of the most compelling polemics from the military revolution belief—that the DoD must search for, invest in, and adopt emerging technologies before its competitors in order to move up the ladder in the historical progression of warfare. This quest shaped the trajectory of unmanned systems in one of the most consistent pushes for technology across many decades and administrations within the DoD.

Tracing the Belief: From Emergence to Cascade to Internalization

Where did these beliefs come from? How did they become a part of the American military narrative? And most importantly for our story, how did they shape the trajectory of unmanned systems in the DoD? Below we trace the military revolution belief from its emergence to cascade and finally to internalization—all the while linking the proliferation of the belief to decisions about investments and uses of unmanned technologies. Finally, while we highlight the role of key policy entrepreneurs and networks of influence in the spread of the belief in military revolutions, we also discuss the role of exogenous shocks like budget cycles and historical events that sometimes pulled these individuals off track—and how they tried to respond to these outside influences.

Emergence

Perhaps counterintuitively, what would become a strongly American view of war originated not within American philosophy but instead from a material and linear understanding of the progression of war latent in Marxist interpretations of history.[15] This is because modern technological determinism debuted not in the United States but instead with the Soviets, who first crafted the military-technical revolution narrative.[16] For the Soviets, military revolutions were discontinuities in the nature of conflict, parallel to the kinds of discontinuities that occurred within societies as revolutions, eventually culminating with communism. Indeed, the use of the term "revolutions" in the belief suggests the underlying influence of revolutions so integral to Soviet thought about history and the world.

Soviet focus on technology as the prime catalyst for revolution also aligned with the materialist focus of Marxist history and explains the genesis of technological determinism within the military revolution narrative. Soviet military strategists, keen to fit military history within a Marxist historical lens, sought out technological impetuses for discontinuities in the history of war. These first Soviet thinkers found two—the mechanization of warfare and the development of nuclear weapons. Drawing lessons from these two previous revolutions, Soviet strategists in the mid-1960s and early 1970s foreshadowed an impending military-technological revolution, catalyzed by automated and long-range technologies that could create range and, eventually, compress detection-destruction cycles.[17] As Adamsky explains, "Particular attention was devoted to automated decision support systems, telecommunications, and enhancing accuracy, range and lethality of stand-off and the direct attack PGMs."[18] It was Marshal Nikolei Ogarkov, chief of the general staff, who, starting in 1977, brought these concepts into popular and widespread use within Soviet strategic dialogue. Ogarkov wrote a series of articles on the military-technical revolution arguing that the next revolutionary advantage would be in the use of new technologies to see and strike deep behind enemy lines.[19]

For years, these writings were translated by US intelligence and played little role in American strategic thought.[20] While technology was a large player in the second offset championed by future secretary of defense William Perry, there was no underlying narrative about technology's agency in creating revolutionary discontinuities in the character or conduct of warfare across a historical progression. In fact, in many ways, the "American" way of warfare leading into Vietnam was one of mass, attrition, and the stubborn will of the American people to persevere regardless of technological shifts. That changed in the late 1980s when Andrew Marshall, director of the small ONA—a strategic initiatives group that reported to the Office of the Secretary of Defense—commissioned a translation and study of the Ogarkov writings on military-technological revolution.[21]

Marshall believed the Soviets were on to something, a hunch that seemed to be validated by the 1982 Israeli War in Lebanon and the resounding success of Israeli high technology tactics in the Beka'a Valley. In particular, Marshall and his team noted the Israeli use of drones and precision-guided munitions as part of an Israeli tactic to seek and destroy targets at longer ranges and in condensed timelines.[22] By the mid-1980s, Marshall began in earnest to focus his staff on fleshing out the Soviet beliefs, determining whether the United States could co-opt the revolution, and investigating what implications there were for future warfare. At the same time, the Army was developing a new doctrine, AirLand Battle, which also drew lessons from Israeli successes with new, long-range technology. AirLand Battle, like the Soviet military-technological revolution, looked at emerging innovations in precision-guided munitions and called for an offensive doctrine that struck deep behind enemy lines and, crucially, introduced time as an important characteristic of future war. The doctrine called for remotely piloted vehicles, long-range standoff precision munitions, surface-to-air missiles and smart rocket systems as part of the arsenal of weapons needed to implement the doctrine.[23] These conceptions of the future of war aligned well with Marshall's predilections about military revolutions and the future of war.[24]

While the idea of military revolutions was increasingly engaging Marshall's attention, the belief in independent technological shocks to military effectiveness was only starting to emerge within the DoD. Marshall's staff and budget were small. He had no authority to acquire technology, invest in research and development, or develop campaign plans or strategy. In Pentagon influence terms, he had almost none of the tools usually required to create a large shift in weapons trajectories.

This is where Marshall's real and enduring genius manifested; he was extraordinarily savvy about devising ways to influence defense thinking absent the Pentagon's normal bludgeons of influence: budget and campaign plans. The first part of the strategy was to create circles of influence from both up-and-coming personnel and established thinkers. This strategy to propagate the military revolutions narrative into mainstream defense thinking was first used in 1986. Marshall (with Freddy Clay) led a working group as part of a commission on long-term thinking. The commission was led by political scientist Albert Wohlstetter, who was an early believer in military technological transformations but whose influence would wane in subsequent generations. Marshall's group was composed of hand-picked members culled by Marshall and Clay. These chosen members were identified as innovative and strategic thinkers, many of them young voices within the Pentagon that included Eliot Cohen, James Roche, Stephen Rosen, and others. The group concluded that "technology . . . is partly exogenous in its development" and stressed the importance of discontinuities to the nature of warfare.[25] The group directly addressed Soviet writings on the military-technological revolution (including them in the appendix) and argued that "the Soviets are correct in their assessment that the advent of new technologies will revolutionize war, and not merely make current forces marginally better at what they do. In the same way that long-range rifles and railroads transformed combat in the mid-19th century (and tanks and aircraft did in the mid-20th century), the new technologies will profoundly alter tactical requirements, operational possibilities, and even, in some cases, strategic choice in the early 21st century."[26] The report also highlighted the role of unmanned technologies to create these revolutionary advances in warfare—in particular, unmanned vehicles as intelligence platforms, smart mines, and cruise missiles as replacements for more expensive manned aircraft. As the report notes,

Unmanned vehicles can encompass everything from remotely piloted vehicles to robotic systems capable of detecting, classifying, and attacking whole families of targets with relative autonomy once they have been set in motion. During the Vietnam war, U.S. forces successfully pioneered the use

of reconnaissance drones; their utility for very high threat environments was reconfirmed during Israel's 1982 campaign in Lebanon (albeit in well known, nearby areas). Looking ahead, advances in sensor technology combined with progress in areas like automatic target recognition and expert reasoning will eventually make possible families of robot-like vehicles able to provide robust complements to manned systems. Possibilities range from remote battlefield fire units and more intelligent submunitions to mine-like aerial drones that can be seeded over enemy airbases and attack aircraft during peak launch and recovery periods....[27]

... Cruise missiles represent the leading edge of standoff, very high accuracy weapons.... Such systems could be useful in holding at risk with conventional munitions targets deep in the theater or deep within the Soviet Union. Even in the near term, cheap cruise missiles, whether launched from aircraft, ships, or army systems like multiple rocket launchers, should be able to achieve almost pinpoint accuracy against any targets whose positions can be precisely located, regardless of time of day or weather conditions. Most of the more lucrative interdiction and FOFA (follow-on forces attack) targets that could be covered with long-range, high accuracy weapons have long been assigned to land-based aircraft. Many of these targets could be covered by conventional cruise missiles, thus freeing up manned aircraft for other important missions.[28]

Marshall's group fed into the larger commission, which included erudite (and influential) thought leaders in the DoD and national security landscape, including Henry Kissinger, Samuel Huntington, and Zbigniew Brezinski, as well as chairs Fred Ikle and Albert Wohlstetter. While only a small offset group of the overall commission, Marshall's group and their beliefs about the military-technological revolution proved influential in the commission's final report. The commission noted that "the Soviet military establishment is already engaged in a major effort to understand the military implications of new technologies, and appears to have concluded that revolutionary changes in the nature of war will result. The much greater precision, range, and destructiveness of weapons could extend war across a much wider geographic area, make war much more rapid and intense, and require entirely new modes of operation."[29] It concluded that the Soviets were likely right, foreshadowing "sharp discontinuities as well as gradual changes with effects that are cumulatively revolutionary."[30] The commissioners warned that microelectronics could create a revolutionary command

and control capability, perhaps so powerful that it could make the previous revolution of nuclear weapons obsolete.

The commission served as a rallying event for the initial cultivation of the military-technological revolution idea, and the work highlighted the importance of unmanned technology within the current revolution. However, despite Marshall's initial forays to influence DoD thinking by introducing the belief to key policy entrepreneurs, the recommendations of the commission were overshadowed by the collapse of the Soviet Union. The Cold War was over and with it the impetus (and budgets) for a massive technological modernization. It could have been the end for military revolutions and unmanned systems. However, Marshall was nowhere near done, nor was the influential role of many of Marshall's fellow working group compatriots, especially in propagating the military revolution narrative and fueling the drive for unmanned technologies. With the demise of the United States' principal adversary, Marshall recognized a strategic vacuum of ideas and increased efforts to insert the military revolution as the primary strategic thought for a post–Cold War United States. In 1991, he turned to his ONA colleague, Andrew Krepinevich, to write an examination of the military-technological revolution. Were the Soviets right? And if so, how could the United States co-opt the idea to leapfrog over future competitors?

Cascade

Krepinevich's 1991 examination of the military-technological revolution was published in 1992 and explained that the Soviets "assert that this revolution's first stage will see far greater reliance placed on rapidly acquiring, processing, and moving surveillance and targeting information than during the Cold War era, thereby increasing the value of space-based systems, unmanned systems, and automated detection and engagement."[31] Krepinevich traced the beginnings of this military revolution to the 1970s, arguing that it was unmanned systems and in particular "the arrival of reliable ballistic and cruise missiles, the deployment of a military satellite network, and the arrival of precision-guided munitions"[32] that ushered in a revolution in warfare.

He identified three takeaways for US defense policy from his evaluation of what he perceived as an impending information technology (IT) revolution. First, Krepinevich argued that the current revolution called for information dominance. Second, this information dominance would be key to winning in wars, which would increasingly be a competition between hiders and finders, in which a geographic forward edge of battle becomes irrelevant. Finally, he argued that if employed effectively, these revolutionary technologies would also be able to conduct war by striking fewer targets and ultimately decreasing the scope of violence in war.

To capitalize on these revolutionary effects, Krepinevich called for new weapons and ways of thinking. He foreshadowed "the use of space platforms, unmanned air-breathing aircraft, high-speed computers, and a variety of sensors to gather, process, and move huge amounts of relevant information"[33] and called for systems that were less visible or observable, could operate beyond line of sight in networked architectures, and were increasingly automated "to gain/maintain information dominance by minimizing the expenditure of time."[34] Krepinevich argued that many existing systems would not compete in this new concept of war and so he warned of sunset systems like heavy tanks, large-surface combatants, satellites, and (most importantly for our story) manned aircraft. As Krepinevich explains, "We are only beginning to tap the potential of long-range munitions (e.g. cruise missiles), unmanned aerial vehicles (UAVs), and their linkage to space systems in what may be a new form of military operation: the aerospace operation. It may be that unmanned systems and standoff munitions, perhaps coupled with U.S.-based strategic stealth aircraft, will constitute the 'tip of the spear' in future conflicts. These systems could disable the enemy's ability to resist attacks by more 'traditional' manned systems."[35]

The examination immediately got traction. Not only was Krepinevich's narrative clearly articulated and compellingly written, but also it had on its side both a bureaucratic push from ONA and the fortune of timing. Bureaucratically, once Marshall was convinced of the merit of the military-technological revolution, he sought to increase the influence of the idea by organizing a series of discussions, seminars, and wargames populated by thought leaders within the services. Marshall knew that the services, which had the money to buy and develop weapons, were the key to the real influence that would normalize the military-technological revolution within the defense establishment. The subsequent proliferation of these weapons and technologies from the budgets proffered by the services would then provide the left- and right-hand limits of the development of combat campaigns. As Rosen (one of Marshall's initial group from the commission on long-term thinking and later professor at Harvard) writes, "Mr. Marshall never stated what he thought the shape of future American military force structures or concepts of operation should be. He regarded this as, intellectually and morally, a decision that the services had to make for themselves. He did, however, wish to facilitate service efforts to think this issue through, and to that end, sponsored a series of seminars with the services."[36] These within-Pentagon circles of influence that were actively promoting the military-technological revolution coincided with the fortune of timing—an important step in propagating the belief and also encouraging the services to participate in Marshall's seminar series.

First, in 1991 (as Krepinevich was developing his work on military-technological revolutions), the United States launched Operation Desert Storm. The war in Iraq unveiled new technologies and a concept of operations dazzlingly effective against what was thought to be a formidable Iraqi force. A combination of massive air support, digital coordination, and intelligence, surveillance, and reconnaissance enabled a quick and decisive victory over one of the largest military forces in the world—all with only losing approximately 150 American forces. In an after-action report examining the (unexpected) success of the war, Dick Cheney, secretary of defense, attributed much to the revolutionary narrative, declaring that the Gulf War "demonstrated drastically the new possibilities of what has been called the 'military-technological revolution in warfare'" in which "precision guided munitions proved immensely effective."[37] While merely a note in the DoD's report to Congress on the conduct of the war, the Pioneer UAV was highlighted as "valuable and appears to have validated the operational employment of UAVs in combat based on preliminary data."[38]

With the end of the Cold War, the defense establishment was eager for new ideas; Krepinevich's examination of the military-technological revolution created a compelling narrative, and the Gulf War seemed to demonstrate the power of the military-technological revolution in practice. Military revolutions seemed to explain the dramatic victory in Iraq, and defense practitioners quickly put the ideas of the military revolution to paper, calling on investments in technologies to capitalize on the new technology revolution. As future secretary of the Air Force and undersecretary of defense for acquisition, technology, and logistics Frank Kendall wrote in the spring of 1992, "Together these technologies provide a quantum improvement in the efficiency with which force can be applied on the battlefield. If a commander can see the entire battlefield with high fidelity, if he can make operational decisions in near real-time, and if he can target and engage in accordance with those decisions with high efficiency weapons, then that commander needs little more." Unmanned systems were a vital part of Kendall's sensing revolution—in terms of both reconnaissance and ultimately strike. In the paper, Kendall highlighted the experimental use of UAVs in the Gulf War and called for their expansion as part of a system of sensors to enable long-range and rapid targeting.[39]

Marshall also attempted to inculcate the belief within the services, organizing sessions of leading thinkers across the services to introduce the belief to those who could influence both doctrine and capabilities development (such as Air Force Colonel John Warden, Admiral William Owens, Frank Kendall, Johnny Foster, and retired generals Meyer, Gorman, and Starry).[40] By the mid-1990s, beliefs in the Soviet-introduced military–technological revolution were cascading across the American defense establishment.[41] The materialist and Marxist undertones of the military-technological revolution had been replaced by a uniquely American RMA. As Dima Adamsky argues, "Not until Andrew Marshall and his colleague from ONA introduced the notion of the RMA into the discourse of the American defense community did this conceptual innovation reach its consciousness."[42]

While the dissolution of the Soviet Union provided an opportune strategic vacuum for the cascade of beliefs about the military revolution, it also introduced major political challenges for proponents of military revolutions beliefs. Whereas Marshall had significant influence with Reagan's outgoing secretary of defense, Dick Cheney, his relationship with Cheney's successor, Les Aspin, and Aspin's head of the Office of the Secretary of Defense Policy Planning Staff, Clark Murdock, was less constructive. As Krepinevich and Watts recount of a meeting between Murdock and Marshall, Murdock wasn't convinced that military revolutions appealed to the service chiefs, who "weren't worried about any major challenge to US military dominance arising anytime soon," and that the military revolution assessment "wasn't going to help 'diddly shit' in Bosnia or Somalia, and that's what people care about now. . . . [T]he MTR didn't do a goddamn thing for marines in Somalia."[43]

While Marshall's relationship with the secretary of defense improved under Bill Perry in 1993 (who had long been an advocate of the importance of technology to strategic success), Perry was not much more receptive to the new military-technological revolution narrative than his predecessor. As Krepinevich and Watts explain, "Perry was also of the view that the military revolution had largely already happened, that it had emerged from the work he had done under Harold Brown on stealth and Assault Breaker during the Carter Administration. Unlike Marshall, who believed the MTR was in its very early stages, Perry believed the opposite."[44] Indeed, Perry's beliefs about the purpose (and genesis) of unmanned systems aligned closely with efforts he took in the second offset to develop satellites and smart weapons, all undergirded by innovations in semiconductor technology. As Perry recounted, "Unmanned systems played a key role in second offset. The second offset strategy was introducing the then emerging semiconductor technology into defense systems and that manifested in many ways, but one of the principal ways was in the development of unmanned systems. I consider the offset strategy to be the smart weapons and unmanned systems are components. Certainly what I had done on unmanned systems and satellite systems hugely influenced the concept of the offset strategy and the way we implemented it."[45] It wasn't so much that Marshall's and Perry's ideas were at odds as that Marshall saw an urgency in the adoption of new technological systems, and unmanned aerial platforms in particular, that Perry viewed more as a continuity within the arsenal of nuclear ballistic and cruise missiles, space reconnaissance platforms, and tactical unmanned systems that dominated the Cold War US unmanned arsenal.

Absent Marshall's influence with the secretary of defense during this time, service chiefs reigned supreme, and investment in the kinds of technologies required to decrease time between the sensor and the shooter (the kill chain accelerators) envisioned by Krepinevich languished. Investments in new systems, not already aligned with core service identities, were especially difficult to get into the budget as the fall of the Soviet Union also led to a significant budget retrenchment. Paradoxically, the term "RMA" continued to lace defense strategies and budget requests. For example, the Army's Training and Doctrine Command (TRADOC) headquarters wrote in 1994 about the information revolution's influence on warfare. They argued that the new revolution increased the "technical means to generate, share, and understand information . . . [as] a result of evolutionary integration to network the force with advanced sensors, processors, and communication devices."[46] In practice, however, the narrative was used to advocate for platforms and programs the services already favored. As Frank Kendall remembers, "In the '80s and '90s I was a proponent of UAVs in the Pentagon and it was a pretty lonely thing. . . . [W]e had a hard time getting budgets through. It was hard to get programs through in that budget environment (after Cold War). It was easy to experiment with things but getting people to actually field things is harder."[47]

This was a difficult time in the growth of the military revolution idea, made more difficult by the fact that Marshall's ONA had no ability to directly control DoD purse strings. However, Marshall's network was growing, cultivated with hand-picked true believers in military revolutions.[48] The true believers had become influential defense strategists in their own right and were now finding ways to keep the military revolution belief alive and reinvigorate investment in the types of technology the group believed would change the nature of war. By 1997 and the Quadrennial Defense Review (QDR), military revolution entrepreneurs like Andrew Krepinevich had made inroads into committees advising on appropriations. While the service chiefs pushed their favorite service capabilities in budget requests (primarily manned platforms), organizations like the National Defense Panel provided an outside push for the military revolution by recommending and prioritizing investments to Congress. That's why, despite ONA's official lack of influence in the late 1990s budgets, the 1997 QDR and National Defense Panel still explicitly called for transformative technologies.[49] This allowed the military revolution to continue to influence capability development in the military, even during Secretary of Defense William Cohen's tenure—a particularly difficult secretary of defense tenure for Andy Marshall's group with ONA ignominiously relegated out of the Office of the Secretary of Defense and hidden under the administrative control of the National Defense University.

While Marshall may have been tucked out of the way of the secretary of defense across the Navy Yard in Fort McNair, he used his relationship with the professional military education establishment wisely. Perhaps most notably, Marshall commissioned two well-respected historians, Williamson Murray and Alan Millett to write three volumes on military effectiveness from World War I to World War II.[50] These volumes would become the academic underpinning of the military revolution narrative (and, conveniently, also replaced Soviet thinkers as the primary authors of military revolution history), serving as primary source documents for subsequent work by Marshall's inner circle and second generations of scholars researching military effectiveness.[51] Marshall vaulted these volumes into prominence within military circles by introducing them to professional military education. Throughout the 1990s, he and his inner circle led a summer study at the Naval War College, thereby increasing both the proliferation of these commissioned works and the network of true believers. By 2020, these three volumes by Millett and Murray had been cited over 1,100 times and the idea of military revolutions had led to dozens of new historical works on military revolutions. As Williamson Murray recounted, Marshall "affected how military historians consider larger patterns of earlier innovation and adaptation."[52] These works became a staple of professional military education curriculum, taught at places like the Naval War College by members of Andy Marshall's inner circle including Steve Rosen, Eliot Cohen, and Tom Mahnken. As these members later migrated to elite civilian universities like Harvard and Johns Hopkins, Marshall's influence followed, inculcating military revolutions in multiple generations of security scholars.

Marshall's influence strategy during this period appeared to be in devising bottom-up efforts to evangelize military revolution beliefs—including the 20XX wargame series in which the Center for Strategic and Budget Analysis (CSBA) recruited emerging thought leaders from all of the services to play in an iterated wargame about the future conduct of war. Organized by Michael Vickers, a former Army special forces officer and a student of Cohen and Krepinevich at Johns Hopkins, the games gave players notional future capabilities and asked them to utilize these capabilities in new and innovative ways to win campaigns. Unmanned systems—and in particular unmanned aircraft—were a large component of the games, and Vickers's paper for ONA on unmanned systems in the future of warfare (partially written under the tutelage of Cohen and Krepinevich at Johns Hopkins' School of Advanced International Studies, SAIS) became a catalyst for the games. The games, played iteratively across a wide swathe of up-and-coming defense thinkers, not only were learning games to brainstorm future warfare but also served as a polemic for these future capabilities (particularly unmanned systems). Players such as Bob Work (who would later work for Krepinevich and Danzig and coin the third offset as deputy secretary of defense), Johnny Foster, Bob Martinage, Tom Erhard, and Don Hicks would all become advocates for unmanned systems over the next twenty years. As Work recounts about the creation of the games and their use as an idea proliferator:

> The first several games of the 20XX series were a bust. We would bring
> in operators from the 4 services and would describe broad trends of the

emerging military-technical revolution and ask them to develop new oper-
ational concepts consistent with the trends. This proved to be too abstract
for the operators who just couldn't come up with new operational con-
cepts. So ONA wrote a playbook—a gamebook that described in detail the
capabilities and characteristics of systems consistent with the tends and
ones specifically designed to play in these regimes (UAVs, stealthy aircraft,
hypersonic weapons, directed energy weapons, undersea projection capa-
bility and unmanned systems across the domains). The players started
to play with these capabilities which in turn suggested the new opera-
tional concepts we were hoping for. Players (maj-lt col, col) went back
to their services and by 2010s many were in leadership positions in doc-
trine and concept commands. I think it is fair to say these games helped
in large measure to make the idea of unmanned systems more palatable
to these operators, as they helped operators gain trust in a future with
unmanned systems. One reason for this, I think, is that unmanned sys-
tems did not replace humans in the games. Instead, they were used in "dull,
dirty, and dangerous" situations. 20XX games socialized unmanned sys-
tems and showed how unmanned systems might work with humans on the
battlefield. .[53]

With these games, academic investments, and network expansions, Marshall kept military revolutions alive through largely hostile (or at least ambivalent) secretaries of defense. As the 1990s came to a close and the Joint Vision 2020 was published in May 2000, military revolutions and the idea of an information-centered transformation in military affairs had survived almost a decade of focus on force protection and irregular conflicts. While the document focused on many of the operations and threats from the decade, it recognized that "If our Armed Forces are to be faster, more lethal, and more precise in 2020 than they are today, we must continue to invest in and develop new military capabilities. This vision describes the ongoing transformation to those new capabilities. As first explained in JV 2010, and dependent upon realizing the potential of the information revolution, today's capabilities for maneuver, strike, logistics, and protection will become dominant maneuver, precision engagement, focused logistics, and full dimensional protection." It continued on to state that the "joint force must be able to take advantage of superior information converted to superior knowledge to achieve 'decision superiority'—better decisions arrived at and implemented faster than an opponent can react, or in a noncombat situation, at a tempo that allows the force to shape the situation or react to changes and accomplish its mission."[54]

Throughout the 1990s, despite Marshall's success in creating prolific influence networks, commissioning thought pieces to support the idea by both academics and think tanks, and ultimately normalizing military revolutions as a belief within the Pentagon, he still struggled to move the needle on capabilities investments. Part of this was the unfortunate timing of the advent of military revolutions in the midst of a major drawdown in military budgets. With the lean years of the 1990s, services fought hard to protect systems that supported their core identities (see Chapter 6), while the strategic focus of the Clinton administration was not on optimizing the American military for peer competition but instead on mitigating risk to US military personnel in a series of humanitarian and periphery conflicts. Unmanned systems, which military revolutions beliefs suggested were key to speeding up the sensor-shooter chain, were instead being invested in only sparingly by the services and usually as an unmanned replacement for an otherwise standard manned mission set. Where they were used, it was not in a novel way to augment networks but instead to mitigate risk and provide persistent surveillance. Despite the fact that the belief in military revolutions was cascading and unmanned systems were increasingly integrated into campaigns in Bosnia or against terrorists, the focus on force protection and incremental advances to legacy systems was still far short of the military revolution that Marshall's coterie envisaged.

Internalization (with Hiccups)

The 1990s may have been a period of careful bottom-up seeding for Marshall, but the 2000s breathed new life into military revolutions from the very highest level of government, now transformed as RMA. After two decades of cultivation, military revolutions were moving beyond the hard-earned cascade of the 1990s to full internalization within the US military. Marshall's early cadre of military revolutions proselytes, who had now left the small Pentagon office, had moved on to positions of influence. They had formed think tanks of their own (often funded by Marshall's office), staffed top universities, and were waiting in the wings for the opportune moment to rejoin the inner circles of defense leaders. In 2000, a survey fielded by Thomas Mahnken and James Fitzsimonds at the Naval War College found that 85% of the officers surveyed in the professional military education course believed in the IT-RMA.[55]

The opportune moment occurred in 2001 when a change in presidential administrations provided a major accelerant to Marshall's influence networks. Even before his election as president, candidate George W. Bush had articulated the military revolution narrative (likely the influence of soon-to-be secretary of defense Donald Rumsfeld), declaring in a campaign speech at the Citadel that "The real goal is to move beyond marginal improvements—to replace existing programs with new technologies and strategies. To use this window of opportunity to skip a generation of technology." This would occur in part by striking "from across the world with pinpoint accuracy—with long-range aircraft and perhaps with unmanned systems."[56] As President Bush started his term in early 2001, he staffed a department to create that transformation, nominating a series of Pentagon old-hands, many with past relationships with Andy Marshall and ONA. Bush Sr.'s former secretary of defense Dick Cheney, previous believer in the role of transformative military technologies in the Gulf War, became vice president. Andy Marshall's assistant director of ONA from 1976, James Roche, became secretary of the Air Force. And Donald Rumsfeld, who as secretary of defense under the Ford administration had first endorsed Marshall's influential net assessments within defense thinking,[57] became secretary of defense (for the second time) in 2001.

Under Rumsfeld, Marshall and ONA quickly regained their influence within the Pentagon. RMAs were back in a big way, and the service's pet capabilities (all manned platforms) were under threat. Most notably, as the 2001 QDR loomed, Rumsfeld asked Marshall to spearhead a strategic study that would guide the final report and ultimately wrestle influence away from what Rumsfeld saw as service chiefs wedded to old ways and weapons of war. As Maddrell recounts, "The Secretary of Defense initially proposed the core study be undertaken by Andrew Marshall, the hub of whose operational universe was not twelve carrier battle groups, the F-22, the Abrams tank33 or Crusader. Marshall was believed to think the Services were not preparing for the next likely theaters of war. There was no reason for them to think his strategic review would do anything other than reiterate that, and possibly include an unflattering assessment of existing states of readiness."[58]

When the 2001 QDR did come out, it heralded the return of military revolutions as the predominant underlying belief system motivating weapons acquisitions programs in the US military. As the report declared:

The ongoing revolution in military affairs could change the conduct of military operations. Technologies for sensors, information processing, precision guidance, and many other areas are rapidly advancing. This poses the danger that states hostile to the United States could significantly enhance their capabilities by integrating widely available off-the-shelf technologies into their weapon systems and armed forces. For the United States, the revolution in military affairs holds the potential to confer enormous advantages and to extend the current period of US military superiority. Exploiting the revolution in military affairs requires not only technological innovation but also development of operational concepts, undertaking organizational adaptations, and training and experimentation to transform a country's military forces.[59]

Interestingly, in the document, which was largely written before 9/11, the focus on unmanned systems within this military revolution was in creating situational awareness and intelligence to enable transformative long-range standoff strike capability. This was a vision of unmanned systems that aligned very closely with the military revolution's understanding of technology's ability to create an IT revolution.

9/11, Iraq, and Afghanistan—The RMA Derailed

Despite the fertile climate for military revolutions within political elites at the top rungs of the Bush administration, an exogenous shock derailed the forward progress of military revolutions and their investments in unmanned systems. On September 11, 2001, terrorists took down the World Trade Center and careened into the Pentagon. Decision-makers, previously focused on a high-tech transformation of the US military to combat a rising China, needed new options for a low-tech Taliban in remote and mountainous Pakistan and Afghanistan. Luckily, the military had an option—one that had been a pet project of a small defense industrial base company and already tested in Bosnia and Kosovo as a persistent source of overhead real-time video. For years, decision-makers had toyed with arming the Predator but balked at equipping the remotely controlled system with munitions. But now, eagerly looking for a platform that could fly in politically fraught airspace for long periods of time, Secretary of the Air Force James Roche (and previous ONA staffer) told the Air Force to arm the system.

The United States had come to a fork in the road for unmanned systems (and weapons investments for the information revolution), and the US military decided to plunge headlong into tactical reconnaissance unmanned systems that were remotely controlled and persistent.

This trajectory dominated most of the early 2000s and 2010s, especially as the focus of the DoD shifted away from peer competitors and toward terrorist groups in the Middle East. Campaigns in Afghanistan and Iraq ushered in huge changes in unmanned systems, in particular the dominance of remotely controlled unmanned aerial systems, and galvanized support (and investment) in these systems. But it was never what the original military revolution thinkers had envisioned. While these thinkers continued to push for transformation within the DoD and many of Marshall's original group played prominent roles in the Bush administration,[60] the ideas espoused in the QDR from 2001 were more often derailed than catalyzed by the realities of fighting and funding two wars in the Middle East.

September 11 was a critical juncture in the proliferation of unmanned systems not only because it shifted the focus from cruise missiles and long-range strike to loitering, remote-controlled unmanned systems with high-precision smart munitions, but also because it cemented that dominance with a huge increase in military budgets. The global war on terror (and a previous decade of economic growth) brought with it the largest increase in defense spending since the Cold War, with significant outlays to both campaigns and weapons investments. As Frank Kendall recounts, the "war on terror changed all that, especially for aerial systems. Progress on tech had happened but when we got in a world where we needed things that needed to be observed for a long time ... we ended up buying large numbers (endless appetite in Iraq and Afghanistan). The features of persistence and loiter, to do it in a world where you didn't have a threat really, the mission was really suited and the technology came together."[61]

There may have been a significant increase in weapons investments during the global war on terror, but not all systems were getting a push from it. Many of the systems envisaged by military revolutions believers struggled to survive the acquisitions valley of death. As Ross and Dombrowski recount in their analysis of transformation and defense industrial base outlays during the Bush administration:

What was labeled "the Long War"—roughly the campaigns in Afghanistan and Iraq plus an alleged "global war on terror"—affected both the pace and

extent of transformation. . . . Military spending increased rapidly and the
Afghan and Iraq campaigns further tested, on a much grander scale, weapons
and systems that first appeared with the Persian Gulf War and in the Balkans
during the 1990s. . . . Broad-based transformation goals may have suffered
as the operational challenges of ongoing campaigns served to focus research
and development (R&D) resources on technologies and systems barely on the
radar screen of early transformation advocates.[62]

Perhaps the greatest example of the way in which 9/11 and the global
war on terror affected the weapons and unmanned systems trajectory was
the Army's Future Combat System (FCS). The FCS was perhaps the most
ideal pursuit of the kinds of weapons optimized for an IT revolution. It
included an integrated system of communications, network capabilities, sensors, manned and unmanned systems, and smart munitions. The new force
structure called for lighter, more mobile manned, unmanned, and robotic
vehicles designed to track and outmaneuver enemies through effective information sharing.[63] The final FCS included proposals for UAVs as well as
small unmanned ground vehicles, intelligent munitions, and unmanned
ground sensors.[64] However, the program ultimately failed—the program's
system-of-systems design called for the most difficult and complex set of
requirements the Army had ever worked with, and a firmly entrenched Army
service culture focused not on how all these things linked together (or the
use of unmanned systems) but instead on the development of new manned
ground vehicles.[65] When the threats to Army vehicles changed (namely
the prominence of improvised explosive devices in Operation Iraqi Freedom), the amount of unmanned systems and sensors required to support
the lightly armored manned vehicles made the concept tactically moot.[66]
Finally, the focus on manned vehicles over the network systems required to
link all the elements of the FCS made unmanned systems largely unusable
as they, unlike the manned components, needed networks to function.[67]

But it wasn't just the services keen on pet platforms or the defense
industrial base ill-equipped for the IT-RMA that derailed the trajectory
for unmanned systems envisaged by the military revolution. It was also
a push from the very top. Coming into the secretary of defense position
in 2006 after five years of failed transformation efforts (and knee deep in
the Iraq and Afghanistan campaigns), Robert Gates had no patience for
either service pet projects (especially the Air Force fifth-generation F-22 program) or future technology investments over what the warfighters needed

(or wanted) at that moment. And what the warfighters in command in Iraq and Afghanistan wanted was more Predators and Reapers—not high-tech systems of autonomous long-range strike munitions. Gates fired the Air Force chief of staff and ordered more investment in armed UAVs. As Gates recounted:

> *In the past I have expressed frustration over the defense bureaucracy's prior-*
> *ities and lack of urgency when it came to the current conflicts—that for too*
> *many in the Pentagon it has been business as usual, as opposed to a wartime*
> *footing and a wartime mentality. When referring to "Next-War-itis," I was not*
> *expressing opposition to thinking about and preparing for the future. It would*
> *be irresponsible not to do so—and the overwhelming majority of people in the*
> *Pentagon, the services, and the defense industry do just that. My point was*
> *simply that we must not be so preoccupied with preparing for future conven-*
> *tional and strategic conflicts that we neglect to provide, both short-term and*
> *long-term, all the capabilities necessary to fight and win conflicts such as we*
> *are in today.*[68]

Moving into the second half of the Obama administration, a push from above from Gates, a glut of spending on two campaigns against insurgents and terrorists in the Middle East, and wartime experimentation transitioned remotely piloted aircraft into the dominant unmanned system for both research and procurement in the US military. Notably to the waning power of military revolution beliefs, the focus was not on using unmanned systems as part of a system of systems to combat peer competitors but instead on mitigating risk to US military personnel and creating persistent and remotely controlled surveillance. The military revolution narrative had called for unmanned technology and surveillance, but these beliefs, even internalized within a rising cadre of military officers, had also been hijacked after 9/11. The 2007 roadmap for unmanned systems highlighted this new linkage between beliefs in military revolution and desire to use unmanned systems for persistence and risk mitigation, declaring, "These unmanned capabilities have helped reduce the complexity and time lag in the 'sen-sor' component of the sensor-to-shooter chain for prosecuting 'actionable intelligence.' Unmanned systems are changing the conduct of military oper-ations in the GWOT by providing unrelenting pursuit combined with the elimination of threats to friendly forces; including injury, capture, or death."[69]

Back on Track: New Bottle, Old Wine

But the belief in military revolutions was not dead. Instead, military revolutions experienced a resurrection with the third offset and a remarkable policy entrepreneur—a former participant in Andy Marshall's 20XX games and a leader at CSBA, deputy secretary of defense Bob Work. As Alice Friend explains in her analysis of civilian support for unmanned technologies, "The rise of the Transformation concept and its inclusion of UAVs in conventional combat operations was a continuation of the civilian alliance with UAV proponents and a trend that would be interrupted by the 9/11 attacks and then taken up again 12 years later as the country tired of the counterterrorism wars of the early 21st century."[70]

Work's third offset strategy doubled down on the independent role of technology as a producer of massive changes in military effectiveness. While some of the second-generation work on military revolutions focused more on operational innovation with technology (instead of just technology), Work's efforts focused explicitly on the driver as technology itself. In fact, when asked, Work explained that he preferred the military-technological revolution terminology over the later RMA.[71] As he noted, "offset strategies are clearly driven by military-technological competitions. One way to gain an advantage under conditions of military-technological parity is to develop another technological offset."[72]

Most importantly for the new brand of military revolution, Work saw autonomy as the primary technology that would generate huge jumps in battlefield effectiveness. After receiving a Defense Science Board report on autonomy, Work saw the third offset as an explicit move from an IT-RMA focus on "network" and sensor-shooter to the autonomous facilitation of network warfare. As he explained,

The animating idea of the 3rd offset was to inject as many autonomous narrow task AI systems into each of the 4 grids of a battle network. We hypothesized that at some point you would reach a tipping point and entire battle networks would begin to operate at a higher speed and be able to operate even when the battle network was under attack. We called these human-machine collaborative battle networks. We thought these would be the leading edge of a new war fighting regime called algorithmic warfare. We then envisioned a new military technical revolution involving robotic warfare—the point at which the proportion of unmanned systems to manned systems shifted in favor of unmanned systems. We anticipated the character of robotic warfare would be much, much different from what we see today. To make this future a reality we would need advances in AI to get improved levels of autonomy, and improved levels of autonomy would allow advances in robotics and unmanned systems and platforms.[73]

The third offset therefore envisioned unmanned systems as increasingly automated sensors, platforms, and munitions designed to increase the range of first-strike capability, increase situational awareness, and—ultimately—decrease decision time between sensor and shooter.

Work had learned about more than just military revolutions in his time with Marshall; he had also learned about how to create narratives and influence. In this, however, Work had an advantage over his former colleague. As deputy secretary of defense, Work could also influence the budget. He knew that he needed a story to convince Congress and the White House to invest in the kinds of technologies that had languished during the last two decades of conflict. He needed a good name for what was, essentially, another RMA. As Work recounted, "We had an internal debate in my staff about whether or not to name the technological strategy. Inside DC if you can get a name and it catches hold and it means something to people it can help things move forward . . . I also thought grounding the strategy in a historical lens was important to do. So this led to our adoption of the term 'third offset strategy.'"[74] Equipped with a name and a solid historical legacy reaching all the way back beyond military revolutions to former secretary of defense Perry and the second offset, Work briefed the strategic underpinnings of the third offset in Washington, DC think tank circles and among members of the White House and Congress. Meanwhile, network of second- and third-generation ONA thinkers began to fill out the strateg with think tank reports and publications in blogs and professional militar education journals.[75]

The strategy, the narrative, and the influence campaign worked. As Alice Friend explains, "With warfighter demand still high, the dedication of UAVs to unconventional operations and counterterrorism seemed to have an endless inertia. And then, a puzzling thing happened: civilian preferences began to shift back toward using UAVs as a combat capability for conventional war. Whereas in 2012 civilian preferences for UAV uses were still largely conditioned on warfighter preferences, by 2016 a significant cadre of civilians preferred to think of UAVs much more expansively, including as a capability freed from its mimicking of manned platforms and its exclusion from more conventional combat scenarios."[76]

Work got his funding from Congress and built organizations deliberately outside of the mainstream service acquisition communities. In particular, Work's Strategic Capabilities Office (SCO) became a catalyst for unmanned systems investments—especially those kinds of systems that didn't have an immediate warfighter application or a service to champion them. The SCO focused on manned and unmanned pairing, swarming artificial intelligence systems, and unmanned systems utilizing machine learning to conduct otherwise high-end manned missions like fighter aircraft dogfighting.[77] Even after Work's tenure ended and the department transitioned to a lethality-focused new secretary of defense, Jim Mattis, Work's influence lived on. Former director of the SCO Will Roper became the assistant secretary of the Air Force and took his focus on autonomy and unmanned systems from the outskirts to the mainstream Air Force. Work's voice could also be heard in the 2018 National Defense Strategy, which echoed the third offset narrative and created a jumping-off point for large-scale unmanned system investment across the services.

Conclusion: Technological Determinism, Military Revolutions, and Unmanned Systems

How did beliefs about technological determinism and military revolutions shape the trajectory of unmanned systems in the US military? Unmanned systems were never the overt focus of military revolutions or technological determinism; instead, they were one of a bundle of technologies that together created the impetus for a great discontinuity (or revolution) in military effectiveness. As such, unmanned systems were not an end in themselves but instead a type of technology that could be used for the information revolution to decrease decision-making time and increase situational awareness. Military revolutions beliefs suggested that unmanned systems would tip the balance in sensor-shooter competition—sometimes by being both the sensor and the shooter, sometimes by providing persistent coverage as a sensor, and sometimes by being the shooter itself. The point was how the unmanned systems could create networks of networks, not that the platform necessarily would replace manned systems or that these systems would insulate leaders (or military personnel) from risk.

Military revolutions created a strategic narrative about the purpose of unmanned systems that should have created a trajectory toward autonomous unmanned systems that spread across domains and missions, linked intimately with networks of munitions and sensors that enabled long-range first strike. There was certainly impetus for this kind of investment in unmanned systems—whether it was top down from defense-wide policies like the QDR or through bottom-up drives for new ways of warfare from midlevel officers with relationships with ONA. Further, the survival of the belief in military revolutions across decades, multiple presidential administrations, and a series of conflicts demonstrates its lasting impact on core beliefs within the US military about the conduct of war.

The Army FCS project was probably the purest manifestation of unmanned systems as envisioned by the military revolution narrative. Unmanned systems within the FCS included UAVs (from tactical to strategic), unmanned sensors and munitions, and unmanned ground vehicles that included both tanks and robotic explosive ordnance disposal vehicles. All these unmanned systems were paired with manned components by networks that enabled faster, more maneuverable tactics—all made possible by revolutionary advances in situational awareness provided by the unmanned systems.

But the FCS failed. Its failure tells a lot about how even a dominant belief can fail to influence the ultimate trajectory of a weapon system. The program, which received great fanfare from the highest echelons of the Army, ultimately failed from the bottom up. A poorly prepared defense industrial base, an acquisition community not ready for system-based acquisition programs, new technological challenges, lack of Army cultural support for many of the components, and ultimately an exogenous shock in two campaigns ill-suited for the FCS all culminated in an expensive program failure.

Meanwhile, the success of the Predator provides a counternarrative to the success of military revolutions in unmanned weapons investments. The Predator and the Reaper—two remotely controlled, armed, long-range tactical UAVs—dominated the United States' unmanned arsenal in the decades after 9/11. But this was never the kind of unmanned system envisioned by Marshall's followers. The systems are expensive, not survivable, and not autonomous. They function often as centrally controlled platforms—hardly the networked system of unmanned systems enabling long-range strike laid out in the military revolution narratives. How strange, therefore, that their success was only possible because the belief in the technological impetus for military revolutions created space for the initial experimentation with unmanned systems. Further, when Marshall's networks of believers in military revolutions were promoted through the ranks, they ultimately found themselves in positions to arm and proliferate the Predator and its successors when 9/11 knocked the possibility of true military transformation off its feet.

Despite the failure of the FCS and the success of the military revolution's nonideal Predator, it is the third offset—Bob Work's old wine in a new bottle—that presented the modern US military with the possibility to modernize the military revolutions belief with a suite of new unmanned systems. The Army FCS that failed in 2009 resurrected (albeit with a new name) and included investments in unmanned ground vehicles and the all-important enabling network that got too little attention in the earlier failed version. As *National Defense* wryly commented, "The Army may get its Future Combat Systems—or something akin to it, and sooner rather than later. It looks like it will be an evolution, not a revolution as conceived."[78] The Navy, meanwhile, embarked on the most expensive and concerted effort to develop unmanned systems of all the services with new programs to develop unmanned surface vehicles, carrier-capable UAVs, and underwater unmanned vehicles—all networked together via the Navy Future Integrated Combat Systems. Even the Air Force, which had focused a significant portion of its unmanned efforts into the non-military-revolution-primed Reaper and Global Hawk, turned to networks of unmanned systems as part of manned-unmanned "teams."[79]

But even as unmanned platforms prospered under the auspice of the third offset, unmanned systems like munitions, communications relays, and satellites did not experience the same kind of unbridled exponential growth. Instead, they experienced a series of stutter-starts as a post–Cold War US military struggled to meet the technological needs of the military revolution in the face of acquisition and budget cost reform. Despite the clear relationship between these systems and the military revolution narrative, Marshall's networks of influence couldn't overcome bureaucratic sclerosis to save the unmanned systems that the services themselves wouldn't prioritize. Ultimately the trajectories of these unmanned systems reflected the power of external catalysts—namely 9/11 and counterinsurgency—which saved many of the tactical-user satellite programs and led to extremely precise (if not highly lethal) airborne munitions. But what came at a cost to this catalyst was upgrades and procurement of munitions better suited for a peer competitor in the information revolution era: ground-based and shipborne conventional strike, robust and resilient communications, and strategic early warning. Cruise missiles, ballistic missiles, and mines largely stagnated as the services turned toward unmanned platforms to substitute for manned missions.

Beyond investments in unmanned systems, this chapter teaches us something important about policy entrepreneurs and the networks of influence required to translate beliefs over time into weapons decisions. Marshall and his ONA never had a budget or a direct line between their office's outputs and weapons procurement decisions. Instead, Marshall created webs of influence that would create change over time. As Krepinevich and Watts recount, Marshall's strategy as a policy entrepreneur was to focus on future capabilities: "Marshall believed that 'most of the things the defense secretary could decide or affect were issues that had to do with future forces and capabilities.'"[80] Marshall didn't seek to change one National Defense Authorization Act, but instead—over generations—to change an entire culture of leaders. He paid close attention to the officers and civilians hired by ONA and used his initial cadre to staff commissions, wargames, professional military education, and ultimately senior positions within the defense establishment. As Krepinevich and Watts note about Marshall's strategy of influence, what he could influence "were the US military's characteristics in the mid to long-term future . . . military officers and leaders who would rise to key command and staff positions in the years ahead . . . the priority that would be given to research into new or more advanced military capabilities, which developmental systems would move forward into production, and at what quantities they would be fielded."[81]

Marshall's long-term strategy of influence allowed the belief in military revolutions to survive multiple decades despite significant external shocks, from the demise of the Soviet Union, to the lean years of 1990s budgets and drawdowns, through unsympathetic administrations focused on small-scale conflicts, to finally surviving the biggest shock of all—9/11—and two decades of defense budget glut focused away from military revolutions. Each of these events created critical junctures, some of which derailed unmanned systems on trajectories completely divorced from military revolutions. But Marshall's networks of influence stayed the course, adapting these trajectories and often co-opting them to ultimately preserve the belief system. In the 1990s, they created bottom-up influence networks while adapting military revolutions ideas to the heavy force protection focus of the Clinton administration. And after 9/11, they used campaigns in Iraq and Afghanistan to experiment with unmanned systems, creating the technology required to finally invest in fully autonomous capabilities when an opportune promotion at the end of the Obama administration would provide Bob Work with a platform to pitch a compelling new offset narrative about technology and military effectiveness.

But the belief in military revolutions wasn't the only influence on the American unmanned trajectory. In the next chapter, we will introduce one of military revolutions' greatest competitors—the belief in American casualty aversion and the subsequent dominance of force protection in post-Vietnam American defense circles. How did this belief system, without a true policy entrepreneur or constructed networks of influence, manage to compete with military revolutions? How did the ideas intertwine with one another, and where do these competing beliefs stand now?

Notes

1. Peter W. Singer, "The Future of War Will Be Robotic," CNN Opinion, last modified February 23, 2015, https://www.cnn.com/2015/02/23/opinion/singer-future-of-war-robotic/index.html.
2. "DARPA and the Evolution of Unmanned," DefenseNews, April 2, 2017, https://www.defensenews.com/video/2017/04/02/darpa-and-the-evolution-of-unmanned-tech/.
3. Anna H. Jackman, "Rhetorics of Possibility and Inevitability in Commercial Drone Tradescapes," Geographica Helvetica 71, no. 1 (2016): 1–6.
4. See e.g. Alvin Toffler and Heidi Toffler, War and Anti-War Survival at the Dawn of the 21st Century (Boston: Little, Brown, 1993); Alvin Toffler and Heidi Toffler, Creating a New Civilization: The Politics of the Third Wave (Atlanta: Turner Publishing, 1995); Alvin Toffler and Heidi Toffler, "Getting Set for the Coming Millennium," The Futurist 29, no. 2 (1995): 10.
5. For a comparative analysis of the rise of the US RMA literature and similar waves in Soviet and Israeli defense circles see Dima Adamsky, The Culture of Military Innovation: The Impact of Cultural Factors on the Revolution in Military Affairs in Russia, the US, and Israel (Stanford, CA: Stanford University Press, 2010); Raphael D. Marcus, "Military Innovation and Tactical Adaptation in the Israel–Hizballah Conflict: The Institutionalization of Lesson-Learning in the IDF," Journal of Strategic Studies 38, no. 4 (2015): 500–552; and Stephen P. Rosen, "The Impact of the Office of Net Assessment on the American Military in the Matter of the Revolution in Military Affairs," in Contemporary Military Innovation: Between Anticipation and Action, ed. Dima Adamsky and Kjell Inge Berga (New York, NY: Routledge, 2012), 51–62.
6. See footnote 16.
7. MacGregor Knox and Williamson Murray, The Dynamics of Military Revolution 1300–2050 (Cambridge: Cambridge University Press, 2001), 7
8. Boot, War Made New, 10.
9. Ibid., 455.
10. Andrew J. Bacevich, "Preserving the Well-Bred Horse," National Interest 37 (1994): 43–49.
11. Toffler and Toffler, Creating a New Civilization, 36. On the influence of the Tofflers in the creation of the military revolution belief, see David Jablonsky, "US Military Doctrine and the Revolution in Military Affairs," Parameters 24, no. 3 (1994): 18–24.
12. Martin Van Creveld, Technology and War (New York, Free Press: 2009), 315.
13. Andrew F. Krepinevich Jr., The Military-Technical Revolution: A Preliminary Assessment (Washington, DC: Center for Strategic and Budgetary Assessments, 2002 [1992]); Nikolai

A. Lomov, ed., *Scientific-Technical Progress and the Revolution in Military Affairs (A Soviet View)*, trans. US Air Force (Washington, DC: Government Printing Office, 1974); Office of Technology Assessment, *New Technology for NATO: Implementing a Follow-on Forces Attack* (Washington, DC: Government Printing Office, 1987); Eliot Cohen, "A Revolution in Warfare," *Foreign Affairs* 75 (March–April 1996): 37–54; Joseph S. Nye Jr. and William A. Owens, "America's Information Edge," *Foreign Affairs* 75 (March–April 1996): 23–25; Eliot A. Cohen, Andrew J. Bacevich, and Michael J. Eisenstadt, *Knives, Tanks, and Missiles: Israel's Security Revolution* (Washington, DC: Washington Institute for Near East Policy, 1998); Colin S. Gray, "Technology as a Dynamic of Defense Transformation," *Defence Studies* 6, no. 1 (2006): 26–51; Metz and Kievet, *The Revolution in Military Affairs*; Bacevich, "Preserving the Well-Bred Horse," 43–49; James R. Blaker, *Understanding the Revolution in Military Affairs: A Guide to America's 21st Century Defense* (Washington, DC: Progressive Policy Institute, 1997); Michael G. Vickers, *Warfare in 2020: A Primer* (Washington, DC: Center for Strategic and Budgetary Assessments, 1996); Michael J. Mazarr, *The Military-Technical Revolution: A Structural Framework* (Washington, DC: Center for Strategic and International Studies, 1993); James R. FitzSimonds and Jan M. Van Tol, "Revolutions in Military Affairs," *Joint Force Quarterly*, no. 4 (Spring 1994): 24–31; Daniel Goure, "Is There a Military-Technical Revolution in America's Future?," *Washington Quarterly* 16, no. 4 (Autumn 1993): 175–192; Jeffrey Cooper, *Another View of the Revolution in Military Affairs* (Carlisle, PA: US Army War College Strategic Studies Institute, 1994); Jeffrey R. Cooper, "Another View of Information Warfare," in *The Information Revolution and National Security*, ed. Stuart J. Schwartzstein (Washington, DC: Center for Strategic and International Studies, 1996), 109–131; Paul Bracken, "The Military After Next," *Washington Quarterly* 16, no. 4 (Autumn 1993): 157–174; Gen. Gordon R. Sullivan and Lt. Col. Anthony M. Coroalles, *The Army in the Information Age* (Carlisle, PA: US Army War College Strategic Studies Institute, March 1995); Martin C. Libicki, *The Mesh and the Net: Speculations on Armed Conflict in a Time of Free Silicon*, McNair Paper no. 28 (Washington, DC: National Defense University, March 1994); John Arquilla and David Ronfeldt, "Cyberwar Is Coming!," *Comparative Strategy* 12, no. 3 (1993): 141–165; Antulio J. Echevarria and John M. Shaw, "The New Military Revolution: Post-Industrial Change," *Parameters* 22, no. 4 (Winter 1992–1993): 70–77; Jablonsky, "U.S. Military Doctrine and the Revolution in Military Affairs"; Gen. Gordon R. Sullivan and Lt. Col. James M. Dubik, *Land Warfare in the 21st Century* (Carlisle, PA: US Army War College Strategic Studies Institute, February 1993); and U.S. Army Training and Doctrine Command, *TRADOC Pamphlet 525-5: Force XXI Operations* (Ft. Monroe, VA: US Army TRADOC, August 1, 1994).

14. Millet and Murray, *Military Effectiveness*, vol. II, 336.
15. Though this linear of understanding can also conform to liberalist views of international order in which Hegel is substituted for Marx. See e.g. Carvin and Williams, *Law, Science, Liberalism, and the American Way of Warfare*.
16. Dima P. Adamsky, "Through the Looking Glass"; Arnold Buchholz, "The Role of the Scientific-Technological Revolution in Marxism-Leninism," *Studies in Soviet Thought* 20, no. 2 (September 1979): 145–164; Erik P. Hoffman, "Review: Soviet Views of 'The Scientific-Technological Revolution,'" *World Politics* 30, no. 4 (July 1978): 615–644.
17. M. G. Popkov, "Metodologicheskii analiz informatsionnykh protsessov v sisteme' chelovek-voennaia tekhnika" (PhD diss., Voenno-Politicheskaia Akademiia, 1983); N. Nechaev, "Voennye sistemy sviazi: tendentsii ikhrazvitiia," *Tekhnika i vooruzheniia* 6 (July): 1986; Gen. Ye. Kolibernov, interview in *KZ*, November 21, 1985. As cited in Adamsky, "Through the Looking Glass."
18. Adamsky, "Through the Looking Glass," 264.
19. Nikolei Ogarkov, "Sovetskaia voennaia nauka," *KZ* (Vsegda v gotovnosti, February 18, 1978), 31–43, 59–67
20. Adamsky, "Through the Looking Glass."
21. Robert R. Tomes, *US Defense Strategy from Vietnam to Operation Iraqi Freedom* (New York: Routledge, 2007); Mie Augier, "Thinking about War and Peace: Andrew Marshall and the Early Development of the Intellectual Foundations for Net Assessment," *Comparative Strategy* 32, no. 1 (2013): 1–17; Eliot Cohen, "A Revolution in Warfare," *Foreign Affairs* 75, no. 2 (1996): 37–54.
22. Andrew Krepinevich in discussion with the authors, March 2020; Bacevich, *The New American Militarism: How Americans Are Seduced by War* (Oxford: Oxford University Press,

2013); Andrew F. Krepinevich, *The Military-Technical Revolution: A Preliminary Assessment* (Washington, DC: Center for Strategic and Budgetary Assessments, 1992).

23. Eric C. Ludvigsen, "Future Combat Systems: A Status Report," *Army Magazine* 41, no. 2 (February 1991): 38–44.

24. Richard Lock-Pullan, "How to Rethink War: Conceptual Innovation and AirLand Battle Doctrine," *Journal of Strategic Studies* 28, no. 4 (2005): 679–702; Aaron Blumenfeld, "AirLand Battle Doctrine: Evolution or Revolution? A Look inside the U.S. Army" (BA diss., Princeton University, 1989).

25. Future Security Environment Working Group, *The Future Security Environment*, October 1988, 1, https://apps.dtic.mil/dtic/tr/fulltext/u2/a528366.pdf.

26. Ibid., 26.

27. Ibid., 27.

28. Ibid., 28.

29. The Commission on Integrated Long-Term Strategy, *Discriminate Deterrence*, January 1988, 8, https://apps.dtic.mil/dtic/tr/fulltext/u2/a277478.pdf.

30. Ibid., 64.

31. Krepinevich, "The Military-Technical Revolution," 6.

32. Ibid., 8.

33. Ibid., 15.

34. Ibid.

35. Ibid., 17

36. Stephen Peter Rosen, "The Impact of the Office of Net Assessment on the American Military in the Matter of the Revolution in Military Affairs," *Journal of Strategic Studies* 33, no. 4 (2010): 480.

37. US Department of Defense, *Conduct of the Persian Gulf Conflict: An Interim Report to Congress*, July 1991, I-2, 7.

38. Ibid., 6–8.

39. Frank Kendall, "Exploiting the Military Technical Revolution: A Concept for Joint Warfare," *Strategic Review* 20, no. 2 (Spring 1992): 25.

40. Andrew Krepinevich and Barry Watts, *The Last Warrior: Andrew Marshall and the Shaping of Modern American Defense Strategy* (New York: Basic Books, 2015), 201.

41. William A. Owens and Ed Offley, *Lifting the Fog of War* (Baltimore, MD: Johns Hopkins University Press, 2001).

42. Adamsky, "Through the Looking Glass," 287.

43. Krepinevich and Watts, *The Last Warrior*, 217.

44. Ibid., 219.

45. Bill Perry in discussion with the authors, April 2020.

46. Training and Doctrine Command (TRADOC), "Force XXI Operations: A Concept for the Evolution of Full Dimensional Operations for the Strategic Army of the Early Twenty-First Century," *TRADOC Pamphlet 525-5*, August 1, 1994.

47. Frank Kendall in discussion with the authors, April 2020.

48. Debra O. Maddrell, *Quiet Transformation: The Role of the Office of the Net Assessment* (Washington, DC: National War College, May 2, 2003).

49. National Defense Panel, *Transforming Defense: National Security in the 21st Century* (Arlington, VA: National Defense Panel, December 1997); Peter Dombrowski and Andrew Ross, "The Revolution in Military Affairs, Transformation, and the Defense Industry," *Security Challenges* 4, no. 4 (Summer 2008): 13–38.

50. Krepinevich and Watts, *The Last Warrior*, 195.

51. See e.g. Steve Rosen's *Winning the Next War*; Thomas Mahnken's *The Limits of Transformation, Uncovering Ways of War*, and *Technology and the American Way of War*; and Colin Gray's *Strategy for Chaos: Revolutions in Military Affairs and the Evidence from History*. For historian takes on military revolutions after Murray and Millet's volume, see Winton and Metz, *The Challenge of Change*; Corum, *The Roots of Blitzkrieg*; Johnson, *Fast Tanks and Heavy Bombers: Innovation in the US Army*; Rogers, *The Military Revolution Debate*; and Parker, *The Military Revolution: Military Innovation and the Rise of the West, 1500–1800*.

52. Williamson Murray, "Contributions of Military Historians," in *Net Assessment and Military Strategy*, ed. Thomas Mahnken (Amherst, NY: Cambria Press, 2020), 148.

53. Bob Work in discussion with the authors, April 2020.

54. US Government Printing Office, *Joint Vision 2020: America's Military: Preparing for Tomorrow*, 2000, accessed December 31, 2023, https://www.hsdl.org/?abstract&did=446826.

55. Thomas G. Mahnken and James R. Fitzsimonds, *The Limits of Transformation: Officer Attitudes toward the Revolution in Military Affairs* (Newport, RI: Naval War College Press, 2003).

56. George W. Bush, "A Period of Consequences," speech, the Citadel, September 23, 1999.

57. Thomas G. Mahnken, "Andrew Marshall: In Memoriam," War on the Rocks, April 8, 2019, https://warontherocks.com/2019/04/andrew-w-marshall-in-memoriam/.

58. Maddrell, *Quiet Transformation*, 10.

59. US Department of Defense, *Quadrennial Defense Review Report* (Washington, DC: Office of the Secretary of Defense, 2001), 6, https://archive.defense.gov/pubs/qdr2001.pdf.

60. Eliot Cohen, author of many of the most influential RMA pieces and one of the initial group of the committee for long-term thinking, served as counselor at the State Department; Krepinevich served on the Defense Policy Board under Gates; and Thomas Mahnken, who had also worked at ONA, served as deputy assistant secretary of defense for planning in policy.

61. Frank Kendall in discussion with the authors, April 2020.

62. Dombrowski and Ross, "The Revolution in Military Affairs, Transformation, and the Defense Industry," 22.

63. Christopher G. Pernin et al, *Lessons from the Army's Future Combat Systems Program* (Santa Monica, CA: RAND Corporation, 2012).

64. Ibid., 34.

65. Ibid., 58.

66. Ibid., 115.

67. Not just to make them more effective, as in the manned vehicle case.

68. Robert Gates, "Speech on National Defense Strategy," September 29, 2008, https://insidedefense.com/document/gates-speech-national-defense-strategy.

69. *Unmanned Systems Roadmap 2007–2032* (Washington, DC: US Department of Defense, Office of the Secretary of Defense, 2007), i, https://www.hsdl.org/?abstract&did=481851.

70. Friend, "Creating Requirements."

71. Bob Work in discussion with the authors, April 2020.

72. Ibid.

73. Ibid.

74. Ibid.

75. Robert Martinage, *Toward a New Offset Strategy: Exploiting U.S. Long Term Advantages to Restore U.S. Global Power Projection Capability* (Washington, DC: Center for Strategic and Budgetary Assessments, 2014); Timothy Walton, "Securing the Third Offset Strategy," *Joint Force Quarterly* 82 (2016): 6–15; James R. McGrath, "Twenty-First Century Information Warfare and the Third Offset Strategy," *Joint Force Quarterly* 82, no. 3 (2016): 16–23; Larry Lewis, *Insights for the Third Offset: Addressing Challenges of Autonomy and Artificial Intelligence in Military Operations* (Arlington, VA: Center for Naval Analyses, 2017).

76. Friend, "Creating Requirements."

77. Ibid.

78. Stew Magnuson, "Future Combat Systems Didn't Truly Die," *National Defense Magazine*, September 2017, https://www.nationaldefensemagazine.org/articles/2017/9/26/future-combat-systems-didnt-truly-die.

79. Many of these unmanned initiatives occurred under the technocratic lead of Work's former mentee, Will Roper.

80. Krepinevich and Watts, *The Last Warrior*, 102.

81. Ibid., 103.

Chapter 4
Force Protection and Casualty Aversion Beliefs

There may be such a thing as a cheap airplane, but there's no such thing as a cheap American pilot.

—*Lieutenant General Kelly Burke*

In Chapter 3, we traced how beliefs in the inevitability of technological progression led to a push for exquisite unmanned weapons that proponents argued would create revolutionary increases in operational effectiveness for the US military. These beliefs found their origin in the late 1970s and 1980s and were then developed, propagated, and inculcated by a small but hugely influential organization sitting in the bowels of the Pentagon, the Office of Net Assessment (ONA). From the lineage of thinkers who passed through the cubicles came a narrative about technology and investment, revolutions in military affairs, and then third offset that became internalized within the Department of Defense (DoD) to survive multiple presidential administrations. The strength of this belief influenced a trajectory of unmanned systems that privileged autonomy, integrated systems, and decreased time between sensor and shooter—all in a competition for first strike.

At the same time that these beliefs about technological determinism and military revolutions were circulating and growing in power within the acquisition and technology communities of the Pentagon, there was another belief that emerged organically and shaped a cadre of officers who started their career in the waning days of Vietnam. These young officers, who went on to develop doctrine, plans, and campaigns for the DoD over the following thirty years, staffed the most powerful offices across the DoD, from the Pentagon to the US Army Training and Doctrine Command (TRADOC) at Fort Eustis to the Air War College at Montgomery. Unlike the technology wonks that focused on tactical success by creating a strategic theory of victory, these officers brought to unmanned development a deep and ingrained belief that

The Hand Behind Unmanned. Jacquelyn Schneider and Julia Macdonald, Oxford University Press.
© Oxford University Press (2025). DOI: 10.1093/9780190064419.003.0005

strategic success was tied not only to tactical success but also to the military's ability to win wars without losing public approval. They pointed to the pivotal role that televised images of body bags and body counts played in decimating public support for the war in Vietnam and linked this effect to the United States' ignominious withdrawal. From this belief came a drive for unmanned technology to protect the US armed forces and mitigate the effect of public casualty aversion. While they built maneuver doctrines and fought campaigns in Iraq and Bosnia that leveraged many of the same technologies advocated by the acquisition wonks hawking the technological determinism narrative, it was always with the underlying belief that these technologies were successful not simply because they created tactical effectiveness, but because they enabled campaigns that could win with a minimal loss of US personnel life.

These beliefs distilled into campaigns after 9/11 translating the lessons of Vietnam into a new use of remotely controlled, persistent-coverage unmanned technology for counterinsurgency (COIN) operations in Iraq and Afghanistan. Whereas 9/11 and COIN diminished the power of the military revolutions narrative, thereby derailing unmanned technology, they cemented beliefs about technology and casualty aversion. A hearts-and-minds approach to ground warfare, combined with a desire to mitigate casualties in warfare, enabled an explosion of unmanned technology on the battlefield focused on limiting the violent effects of warfare not only for Americans but also for civilians and even adversaries. This is somewhat remarkable given that previous wars had created an impetus for autonomous technologies designed not to avoid enemy casualties (or even civilian casualties) but instead to inflict as much damage as possible without harming friendly forces. Whereas previously unmanned systems—mines, large warhead missiles and bombs, and even kamikaze drones—had been designed to create advantages in wars of attrition, these new systems were focused on avoiding opting into or getting stuck in those very wars. This would be a remarkably sticky trajectory for unmanned use, in which remotely piloted, large airframes dominated the battlefield. And while the COIN doctrine was far removed from the way warfare was fought in the early 1990s, risk mitigation and public approval remained forefront in commanders' (and foreign policy decision-makers') minds. Surveys conducted in 2015,[1] as well as behavioral analyses of President Bush's and President Obama's decision-making on unmanned systems,[2] showed that mitigating risk to human life was the primary consideration for the US public and drove

US leaders' (especially Obama's) decisions to invest in and use unmanned systems.

In this chapter, we investigate what the casualty aversion and force protection belief is, what kinds of unmanned technology it leads to, and what actors, institutions, and doctrines promote these beliefs. We look at the origin of the belief, from Vietnam and later the Army AirLand Battle doctrine to Weinberger, Powell and Bosnia, and then the intermingling of military revolutions in the Bush transformation years. Finally, we trace force protection beliefs from these initial roots to COIN and the current use of unmanned systems in combat as well as the service's current investments in unmanned systems.

Introduction to the Force Protection and Casualty Aversion Belief

Before we can trace the role of this belief to decisions about unmanned systems, it is helpful to understand what the casualty aversion belief is as well as its relationship with force protection (and spend a brief amount of time looking at whether this casualty aversion exists). The idea of modern American casualty aversion originated to explain patterns of public support for the Vietnam War. However, citizens' willingness to suffer in war and its impact on public opinion for the use of the military is hardly unique to Vietnam. In democracies especially, where leaders have powerful institutional incentives to pay attention to the opinions of their voters,[3] the public willingness to pay the costs of war is a central mechanism by which public opinion can affect foreign policy decision-making. All things being equal, the public wants to avoid its soldiers dying on the battlefield, and yet this is always a risk when leaders use military force. For these reasons a number of key studies have found that when considering whether to use force, leaders in democratic states pay close attention to public opinion.[4]

The early wave of literature on public opinion and support for war was strongly shaped by studies that showed low levels of public support for US military intervention in the Vietnam War. In particular, John Mueller's seminal study of public opinion during the Korean and Vietnam Wars found that support for war decreased in proportion to the log of casualties over time.[5] This finding was later supported by Mueller's analysis of public opinion during the Gulf War in which he argued that public support for the US use of force was much more precarious than many believed.[6]

While Mueller's arguments have been subjected to continued scrutiny and refinement,[7] it is now accepted wisdom that under most conditions, aggregate support for foreign military operations declines as combat casualties rise.[8] More recent studies have shown that battlefield deaths also affect the public's perceptions of a war's progress and its likelihood of success,[9] presidential approval rates,[10] and even the tenure of elected leaders.[11] Taken collectively, these findings became central to institutional explanations for democratic peace.[12] According to Feaver and Gelpi, "Implicit in the argument that democracies behave differently with regard to the use of force is the belief that democracies are sensitive to public opinion and public opinion is sensitive to the human costs of war."[13]

However, when we reference casualty aversion beliefs' impact on investment in unmanned technology, what we are describing is not just whether or not the American public is truly casualty averse (indeed, a more recent wave of scholarship suggests that American support for military losses is much more nuanced than the casualty aversion belief would suggest).[14] Instead, the belief in casualty aversion subsumes a more complex causal story in which DoD decision-makers actually rely on four interconnected beliefs to create the ideational foundation of force protection strategies (that is, strategies that base success on mitigating troop losses). These beliefs are, first, that Americans are casualty averse; second, that these beliefs are strong enough that they catalyze Americans to punish policy makers for military campaigns with significant losses (primarily by forcing them out of political office); third, that policy makers respond to these threats by choosing military campaigns that are less likely to lead to personnel deaths; and, fourth, that this lack of support from policy makers ultimately results in strategic defeat by leading either to surrender or to the use of military campaigns that are suboptimal. Together, these four beliefs lead to investments in unmanned systems optimized for force protection—designed specifically to avoid human losses versus other objectives like taking or holding territory.

Tracing Modern Casualty Aversion Beliefs

Where did these beliefs emerge from? Public opinion about friendly casualties has influenced US decision-makers since the advent of the American military with evidence of American generals and senior defense leaders manipulating (or at least carefully disclosing) casualty costs to the public all the way back to Pershing and World War I.[15] However, our story begins with Vietnam and a generation of young officers that cut their teeth on a war that saw both the loss of their fighting brethren and, ultimately, a strategic defeat. Over a decade of fighting in Vietnam, these young officers observed policy makers who changed operations and tactics to blunt the edge of these casualties within the public eye—and yet never made a strong argument for their necessity, leaving officers to shoulder the burden of loss without the reassurance or hope of strategic purpose. Ultimately, the lesson these officers took from Vietnam was that public sensitivity to loss leads American policy makers to choose technologies and operations that may insulate them from the political ramifications of body bags on the nightly news, but potentially at the cost of strategic success.[16] These young officers, coming of age at the advent of the microprocessor and major computer advances, couldn't influence politicians, but they saw in technology an opportunity to substitute for human risk, potentially without sacrificing battlefield effectiveness.

This created one of the strongest trajectories for unmanned systems as substitutions for manned platforms that increased the physical distance between the human and the battlefield, while still maintaining control and precision. This wasn't about using unmanned systems to create quicker engagements, to condense the sensor-shooter timeline, but instead to provide humans on the battlefield with resources for force protection. This sometimes led to unmanned systems with significant logistical tales, limited automation, and high overall cost. However, these were acceptable trade-offs for systems that could potentially limit the scope of violence in conflict.

Belief Emergence: Vietnam

It is impossible to explain the power of these beliefs in force protection and casualty aversion without first understanding the countervailing beliefs about the American way of war that these young officers leaving Vietnam were rejecting. The generation of generals that led Vietnam campaigns, leaders like Abrams and Westmoreland, had a fundamentally different belief about how America won wars. They saw US victories in Europe, Japan, and to some extent Korea and believed that the American way of winning war was to demonstrate the will of the American people to create and sustain enemy loss at a higher rate than friendly casualties. Victory was about winning in wars of attrition.[17] As historian Steven Casey writes, "Abrams and his team were heirs to a military tradition that viewed relentless campaigns as the best way of keeping down the final American death toll. It was a tradition with a long and illustrious heritage."[18] For these leaders, friendly loss was an intervening variable in the ultimate strategic success, but not the primary determinant. For leaders applying this theory of American war to Vietnam, victory would occur when high-risk search-and-destroy operations, focused on rooting out and killing enemies in a total war, convinced the Viet Cong to completely capitulate.[19] Technology's primary role was not to defend the force (or to create maneuver advantages) but instead to create mass amounts of firepower to increase the ratio of enemy to friendly loss. The United States would win against the North Vietnamese because the United States, aided by B-52s dropping tons of firepower, could outbleed the insurgent force (especially given the lower attrition rates of B-52s versus infantry search-and-destroy teams).[20] There was an underlying assumption for this strategy that the American population would support those losses as long as American body bags came back less frequently than the North Vietnamese.

Abrams and Westmoreland may have believed in the strategic advantage of attrition campaigns over time, but the politicians defending their decision to commit American personnel lives in Vietnam were not as convinced that the United States could stomach the friendly losses required to win a war of attrition with the North Vietnamese. As early as 1965, Johnson was seeking technological solutions to high-attrition ground campaigns from his secretary of defense, resulting in what would become the antecedents to the remotely controlled systems of today: "I don't think there's anyway, Bob, that through your small planes or helicopters ... you could spot these people and then radio back and let the planes come in and bomb the hell out of them?" McNamara's answer to Johnson's query was investment in a weapons program that included anti-infiltration barriers of mines, fencing, and barbed wire—all monitored by a series of unmanned sensors and manned patrols. While the barrier had tactical successes, it couldn't change strategic outcomes,[21] but this didn't stop policy makers from searching for technological substitutes to status quo attrition strategies. As the Vietnam War waged on and Nixon struggled with public opinion about the loss of American lives, his cabinet increasingly looked for technology to substitute for American lives.[22]

The push for unmanned systems focused on mitigating casualties was not just top down. Bottom-up experimentation with unmanned systems began in earnest in Vietnam, even as ground-based attrition campaigns dominated the early parts of the war. Unmanned aircraft flown as decoys and jammers, early laser-guided and inertial navigation bombs, and antiradiation homing missiles and infrared-guided air-to-air missiles all debuted in Vietnam as technological solutions to tactical problems. These experimentation efforts aimed to decrease the risk of casualties from dull, dirty, or dangerous tasks and to provide a tactical edge against Viet Cong high-technology asymmetric tactics.

For instance, in the early 1960s, future secretary of defense Bill Perry led the development of an unmanned intelligence, surveillance, and reconnaissance collection outpost for the US Army that eventually evolved into an unmanned aircraft Guardrail program. The unmanned system, requested by the Army and the National Security Agency, was developed to solve a fundamental problem for their intelligence collection efforts; manned listening posts were dangerous and the Army was losing too many of their intelligence officers. An unmanned listening outpost could substitute equipment for personnel, an inherently less risky choice.[23] This experimentation with unmanned systems in Vietnam was foundational for secretary of defense Bill Perry and influenced his later support for unmanned technologies in the late 1990s. Similarly, Jim Roche, who went on to become secretary of the Air Force and ultimately armed the MQ-1 Predator, first interacted with unmanned systems in Vietnam with an experimental use of a remote-relay unmanned helicopter and sensor to conduct intelligence, surveillance, and reconnaissance along the Vietnam coast. This memory of the utility of unmanned systems stuck with Roche, as did his experiences as a combat search-and-rescue helicopter pilot. As he noted, "I spent too much time in Vietnam picking up downed pilots and crying at night because I couldn't pick up guys that I wanted to pick up."[24] The experiences of Perry and Roche demonstrate the complicated relationship between force protection beliefs, formed from personal experience in Vietnam, and later military revolution beliefs that emerged not through experience but through calculation. Both leaders reconciled military revolutions and transformative technology with force protection, aligning these beliefs to advocate for unmanned technology.

Johnson's early requests for unmanned alternatives to mitigate losses in Vietnam were stymied by the technological limitations of the time. Most notably, the early stage of microprocessor technology made the computing power required for many remote systems too physically heavy and unreliable for significant advances. However, buttressed by bottom-up experimentation happening in niches of the Army and Navy, as well as significant advancements in air capabilities, by 1969, policy makers had increasingly long-range options to employ in Vietnam. These were extremely luring for policy makers looking for battlefield wins without bad press. In response to concerns about friendly losses and effects on public opinion, Kissinger recommended that instead of large ground troop operations, "artillery and air will play a dominant role in these operations."[25]

This pressure for increasingly remote technology to substitute for personnel losses peaked under Nixon.[26] By 1972, with public polls increasingly suggesting the American public was more likely to support bombing over ground campaigns,[27] technology became a key component of Nixon's casualty-sensitive strategy. B-52s equipped with the first high-precision bombs were now able to execute bombing runs from increasingly safe high altitudes. Meanwhile, instead of launching large attrition ground movements to cut off supply routes, the United States was dropping mines in Haiphong Harbor.[28] It wasn't enough. In the end these technological advancements that mitigated risk to friendlies and took the American GI further from the battlefield were not sufficient to either shore up American public support or win the war. As Krepinevich concludes, "U.S. withdrawal from the ground war against the insurgents could have been speeded up by accepting higher casualty rates, but this would have had to be weighed against the accelerated erosion in U.S. popular support for the war effort."[29]

Westmoreland and his generals blamed the public. They were especially conscious of the graphic effects of American losses of life in Vietnam campaigns on the nightly news. Vietnam was the first major conflict to be covered on television, and the introduction of the visual medium brought the American public into the realities of combat while sitting in the comfort of their living rooms. Short news segments and visual tableaus were an evocative replacement for newspapers that might have been more likely to cover entire campaigns and feature far less imagery than television reporting. American military officers were especially wary of how the television coverage of Vietnam affected the American public and blamed the television reporting for much of the loss of public support for the war in Vietnam.[30] In a survey of US generals conducted during the Vietnam War, 89% viewed the press and television coverage as negative; in particular, they viewed the media as "not a good thing, since there was a tendency to go for the sensation, which was counterproductive to the war effort."[31]

Westmoreland repeatedly blamed the media for slanted coverage and, ultimately, for contributing to the failure of US strategies in Vietnam. For example, decades after the Vietnam War, Westmoreland continued to blame Walter Cronkite's pessimistic take on the Tet Offensive for crippling public support and influencing the Johnson administration to adopt less optimal strategies.[32]

Military decision-makers' concerns about television and the erosion of public support for campaigns in Vietnam echoed within the highest level of US decision-makers.[33] Johnson, for example, believed that the television coverage of Vietnam put unique pressure on the executive branch, making Vietnam more difficult to win than the similarly casualty-laden wars in Korea or World War II (a belief he voiced publicly). As he recounted:

> *I thought of the many times each week when television brings the war into the American home. No one can say exactly what effect those vivid scenes have on American opinion. Historians must only guess at the effect that television would have had during earlier conflicts on the future of this Nation: during the Korean war, for example, at that time when our forces were pushed back there to Pusan; or World War II, the Battle of the Bulge, or when our men were slugging it out in Europe or when most of our Air Force was shot down that day in June 1942 off Australia.*[34]

Whether or not the antiwar movement actually impacted the ultimate outcome in Vietnam,[35] the US military largely believed that it did. The belief that public casualty aversion made it more difficult to win wars led to a new theory of the American way of war contingent on popular support.[36] As retired US Army General Hamilton Howze opined, "The record was good until America itself lost much of its will to fight and the politicians and press began their program of vilification."[37]

While unmanned systems played mostly a tactical level role in the overall execution of Vietnam campaigns, US experiences in Vietnam created the impetus for one of the strongest trajectories of unmanned investment. The visceral experiences of Vietnam—both strategic defeat and the very personal loss of fighting brethren—led to a strong belief within the officer corps (particularly the Army) that unmanned systems could stop senseless loss of life. At the same time, many of these young officers had seen the technical ability for unmanned systems to both mitigate risk and perform the dull, dirty, or dangerous missions that had taken the lives of so many during Vietnam.

The generation of young officers who watched a defeat in Vietnam after the ultimately ineffective strategies of attrition rejected the lessons leaders like Westmoreland or Abrams had brought from attrition campaigns in Korea or during World War II. Lives were expensive. Technology could make up for that strategic cost.[38]

Belief Cascade: Reinventing the Military after Vietnam

In the years after Vietnam, the US military (and particularly the Army) needed to reinvent itself. This was an important period for casualty aversion beliefs to emerge and influence military strategy, doctrine, and concepts of operations. The time between leaving Vietnam and the Gulf War included a series of critical junctures for doctrine and weapons acquisition, as well as major exogenous shocks including the demise of the Soviet Union and entry into the Gulf War. Doctrines created by the Army and the Air Force in the 1980s, coupled with the last major acquisition splurge of the Cold War, created a trajectory for weapons designed for quick, overwhelming, and low-casualty wars. For the leaders first developing the doctrine and then leading the military in Desert Storm, the beliefs that emerged from the ashes of Vietnam—US public casualty aversion and the need to avoid wars of attrition—were now codified within doctrine, linked to weapons acquisition, and proven in the Gulf War. Investment in unmanned systems followed these beliefs, with systems that focused on mitigating risk, creating distance between US operators and battlefield threats, and creating persistent intelligence for campaigns predicated on overmatch and swift victory.

The belief's prominence in doctrine begins in the mid-1970s at an Army installation in the suburbs of Virginia where William DePuy led the Army's first major doctrine rewrite since Vietnam. For DePuy and the officers of TRADOC tasked with the doctrine rewrite, the legacy of Vietnam casualty aversion loomed large. They needed an Army doctrine that would insulate the public from significant loss of life, thus making it more likely that the Army could achieve strategic success.[39] But this was harder than just abandoning Vietnam-era attrition warfare. DePuy and his staff drew heavily from lessons learned in the Arab-Israeli wars. Like Andy Marshall, they watched the advent of long-range, precision-strike, and antitank missiles and concluded that technology would create a large-scale advantage for those states able to harness the new technologies. Unlike Marshall, who saw offensive advantage and technology designed to create quick campaigns of technological overmatch, DePuy believed the new technologies would heavily advantage defense by imposing lethal cost on attackers.

These technological advances, combined with a post-Vietnam imperative to limit casualties, led to an extremely pessimistic conclusion in the doctrine. Future wars would not be able to limit casualties and, therefore, the US government would struggle to retain public support after the first major battle.[40] "Active Defense," as the doctrine was colloquially known, recognized the increasing importance of technology to warfare but (similar to Nixon administration tactics with mass airpower) saw the main impact of technology in creating firepower and lethality. Based on these conclusions, the doctrine recommended retrenching and focusing on technological advances to aid defense. As the doctrine explained, "The war in the Middle East in 1973 may well portend the nature of modern battle. Arabs and Israelis were armed with the latest weapons, and the conflict approached a destructiveness once attributed only to nuclear arms. Use of aircraft for close support . . . was greatly reduced by advancing surface-to-air missiles and air defense guns."[41]

Active Defense drew a quick and vociferous response. And it wasn't positive. Early and midcareer officers had no patience for a doctrine that promised the same kind of stymied campaigns of technological firepower that had sunk them in Vietnam.[42] In the five years after Active Defense debuted, over half of the articles in military review journals criticized the doctrine for ignoring the importance of public will and casualty aversion.[43] Young officers, coming from lessons learned in Vietnam and leading a now all-volunteer force, were convinced that the only way to ensure public support was to design a doctrine that enabled short and overwhelming conflicts that mitigated loss of American lives.[44] They saw technology not primarily as a means to create firepower and lethality but instead as a way to enable these types of high-maneuver, offensively dominated campaigns. It should be no surprise, based on this concerted disapproval for Active Defense, that the doctrine lasted only one assignment cycle. With the TRADOC leadership transition to General Don Starry, the Army ditched attrition strategies and firepower technology and instead adopted a new doctrine that also turned to technology, but this time to launch first, from long ranges, and with precision—all to decrease the potential loss of US lives and ensure public support.

AirLand Battle took the beliefs from Vietnam and applied new technology to create campaigns that decreased casualties by striking first and winning quickly. Like the beliefs that were concurrently being developed by ONA on military revolutions, AirLand Battle drew heavily from historical examples of warfare, building a narrative on a historical imperative for technologically-enabled offensive operations.[45] Perhaps most interestingly for our story, AirLand Battle was the first real call for unmanned aerial vehicles within a service-wide strategy. This is especially important for the trajectory of unmanned systems because AirLand Battle was accompanied by the largest acquisition program that occurred before the end of the Cold War (blessed by the large defense budget years of the Reagan administration).[46] The Corps 86 program, as AirLand Battle's acquisition initiative was named, called for remotely piloted vehicles, tactical satellites, long-range artillery fires, medium-range ballistic missiles, and cruise missiles and mines, all linked together by network and sensor capabilities. As such, AirLand Battle promised to provide the doctrinal transition between the experimentation with unmanned systems in Vietnam and a theory of warfare in the late 1980s and early 1990s. It also provided the ideational foundation for the Weinberger and later Powell doctrine.

The Army wasn't the only service interested in unmanned systems after Vietnam, nor was the offensively focused AirLand Battle and Corps 86 acquisition program the only impetus for unmanned investment. US tangential (but deadly) involvement in the Arab-Israeli wars convinced US decision-makers that they needed a way to posture forces without needlessly exposing them to harm. Unmanned systems, particularly unmanned aerial systems, were an appealing alternative. The catalyst, and what made the Navy jump into unmanned aerial systems, was Lebanon. While the United States had left Vietnam, it remained engaged in conflicts in the Middle East—to include a small contingent of Marines on the ground and Navy vessels at sea engaged as part of a multinational coalition involved in peacekeeping efforts in Lebanon. Reagan's decision to send these forces into Lebanon was met with mixed approval—from both his cabinet and the public.[47] It was a politically fraught decision, and one that seemed politically calamitous after 241 Marines were killed by suicide bomber attacks on the Marine barracks in Beirut. When a few months later during a routine air patrol Syrians shot down three Navy aircraft, killing two pilots and capturing one, decision-makers were desperate for an unmanned alternative. As Alice Hunt Friend details in her study of unmanned investment within the US military, "The incident . . . appalled both the Secretary of the Navy, John Lehman, and the Commandant of the Marine Corps P.X. Kelley. . . . After the losses of the pilots in December, both Kelley and the Naval officers recommended to Secretary Lehman that the Navy acquire the Israeli UAV system, and Lehman agreed."[48] The event also heavily influenced then secretary of defense Caspar Weinberger. As *New York Times* coverage at the time recounted, "In his opening remarks Mr. Weinberger said he had been impressed with the pilotless planes, often called drones, when the Israeli armed forces demonstrated their uses while he was visiting Israel after the war in Lebanon. . . . After two United States Navy planes were shot down and a third damaged in a raid on Syrian missile sites in Lebanon last December, critics asked why the Navy had not used drones to seek data on Syrian defenses. The purchase from Israel was apparently made after that."[49]

This vignette reveals how US loss of life created the impetus for investment in unmanned systems that prioritized the mitigation of risk over other characteristics. While the Israelis had used the unmanned aerial vehicles for early warning and detection within manned suppression of enemy air defense packages, Weinberger and the Navy bought the systems chiefly as a replacement for otherwise risky manned reconnaissance missions. These unmanned systems would not be used in the integrated sensor-shooter tactics envisaged by the Israelis but instead to increase persistent surveillance as a force protection measure for ground and naval troops. They were meant— explicitly—to decrease the chance of losing a human pilot. As the DoD's research director, Donald Hicks, was quoted explaining, "The United States can no longer afford to risk so many lives and airplanes against increasingly lethal Soviet defenses and must put more emphasis on unmanned weapons." Hicks lamented that if they had devoted unmanned systems instead of manned pilots to the missions in Lebanon, "smart weapons could have spared men and machines." In the same article, retired Lieutenant General Kelly H. Burke, the Air Force research and development chief from 1979 to 1982, made the argument even more clearly: "There may be such a thing as a cheap airplane, but there's no such thing as a cheap American pilot."[50]

In 1984, shortly after the devastating loss of the Marines in Lebanon, then secretary of defense Weinberger unveiled a new doctrine. The Weinberger doctrine detailed six criteria for the use of the American military, including that (1) the conflict involve "vital national interests of the United States," (2) troops be "committed wholeheartedly and with the clear intention of winning," (3) that there were "clearly defined political and military objectives," which then led to (4) adjustments to the size and composition of forces to make sure that the military could accomplish those objectives, and that (5) "US troops should not be committed to battle without a 'reasonable assurance' of the support of US public opinion and Congress." Together, all these criteria would lead to (6) the use of US troops only as a last resort.[51] As Andrew Krepinevich explained, the conditions of Weinberger's doctrine "represent the army's lessons from Vietnam, and they signify the service's determination to ensure that for it there will be 'no more Vietnams.'. . . [T]here was also bitterness toward the American public for its lack of support or appreciation for the sacrifices made by the nation's men and women in uniform."[52] The Weinberger doctrine codified the beliefs created by Vietnam into a military directive. The need to maintain public support to ensure strategic success was now an official tenet of the US military.

Belief Internalization: The Gulf War, Shrinking Budgets, and a Strategic Vacuum

The beliefs that had emerged from the failure in Vietnam would further cascade in the mid-1980s as the young officers of Vietnam stepped into senior leadership roles within the DoD. Their rise heralded with it doctrinal shifts, but most importantly, their beliefs about the lessons of Vietnam played prominent roles in the campaigns of the 1990s, from the Gulf War and beyond. As Anthony Zinni, a Marine general who would go on to play an integral role in the Gulf War, Somalia, and Bosnia, explained, "The Weinberger Doctrine I think was clearly drawn from the lessons of Vietnam . . . in getting the commitment of the American people behind whatever you do. You aren't the king. You're an elected representative and without the will of the American people which is one of our strengths, but also one of our vulnerabilities. It is the center of gravity for us at a strategic level. And bad guys know that now."[53] In the mid-1980s, Zinni was leading an influential naval futurist group, the Chief of Naval Operations Strategic Studies Group, and followed that assignment as first a regimental commander and then a commanding officer of a marine expeditionary unit. He also commissioned one of the first Marine unmanned support units, explaining, "I saw the reality of this machine in Desert Storm and many other places. This machine served a purpose for our Marines, saving lives and making us more effective on the battlefield."[54]

But Zinni wasn't the only influential officer that vaulted to positions of influence and command from 1985 to 1991. In a 1987 *New York Times* article, Richard Halloran noted that a major change of guard had occurred, placing officers who had spent their formative early years in Vietnam in charge of key components of the US Army. These officers were led by a young new chief of staff of the Army, Carl Vuono, and were determined not to repeat the mistakes of their predecessors—which they saw chiefly as a deficit in the political and public support they needed to win conflict.[55] As Halloran quoted one colonel:

"You have to understand the thinking of those of us who were in Vietnam," *said one Army colonel. Speaking as if he were at a briefing in the White House,* *he said: "Mr. President, don't send us to war again unless you have clear-cut* *political goals and attainable military objectives. Sir, don't send us unless you* *give us sufficient forces and the freedom of action to use them according to* *the principles of war. And, Mr. President, you'd better have a hell of a lot of* *public support."*[56]

This generation of officers spoke candidly about the influence of Vietnam on their beliefs about war and strategic success.[57] They were reticent to support conflicts that they believed had limited political support. With a Soviet threat mostly at bay, these officers spent most of the late 1980s advising against US small-scale conflicts across the globe. As Halloran wrote, "In Vietnam, Army officers believe they fought a war under many handicaps: ill-defined political objectives, negative press and television coverage and waning public support. Today, the military leadership is wary of using armed might as an instrument of national policy, fearing they may again be abandoned to fight alone. For instance, the Joint Chiefs of Staff have cautioned the Administration against becoming militarily involved in the Persian Gulf, opposed the dispatch of marines to Lebanon in 1983, and are counseling against sending combat forces into hostilities in Central America."[58]

Vuono took over as chief of staff of the Army in 1987 and played a major role in combat planning for the 1991 Gulf War against Iraq. However, it was another Army general that led beliefs about casualty aversion to cascade throughout the US military—Colin Powell. Powell became the chairman of the Joint Chiefs of Staff in 1989, an auspicious time to influence the US military. Not only did the Soviet Union collapse shortly after Powell began his term, but also he came to the position during a key shift in power within the DoD. In 1986, the US Congress passed the Goldwater Nichols Act, which centralized the power of the chairman of the Joint Chiefs of Staff as the primary advisor to the president. His advice, therefore, trumped the service chiefs and would serve as the primary voice of advice to the National Security Council and the president. In a unique way in American military history, Powell's beliefs mattered. And for Powell, the belief in American public casualty aversion was strong. While Powell did not go as far as Weinberger in publicly espousing a doctrine of restraint and focus on public will, his beliefs were closely aligned.[59] As Powell explained in his memoir, "War should be the politics of last resort. And when we go to war, we should have a purpose that our people understand and support; we should mobilize the country's resources to fulfill that mission and then go in to win. In Vietnam, we had entered into a half-hearted half-war, with much of the nation opposed or indifferent, while a small fraction carried the burden."[60]

Powell and Vuono loom large as pivotal characters in what was the first major conflict since Vietnam: the Gulf War. The 1991 campaign was led by a cadre of officers who had begun their service in Vietnam and were now reaching their apex with the first unveiling of America's late 1980s technology investment in precision warfare. These officers were loath to repeat the mistakes of Vietnam, and at the forefront of their memory was the impact that public demonstrations and loss of support had made in suboptimal military campaign strategies. General Vuono, at the time the chief of staff of the Army, was leery of the impact of the massive deployment to Iraq on public support. As Woodward details in the days leading up to the invasion n Kuwait, Vuono told his staff, "The big question is political will."[61]

Vuono wasn't the only one worried about political will for the campaign. General Norman Schwarzkopf, who led the campaign, also factored public will high into his campaign design. As they debated the initial campaign design, Schwarzkopf even considered an airpower-only effort, reflecting the chief of staff of the Air Force General Dugan's perspective that "air power is the only answer that's available to our country to avoid a bloody land war."[62] (Dugan was fired by the Bush administration for his willingness to go straight to the public to argue for an airpower campaign.) Congress also warned the Bush administration that whatever military option they chose in Iraq had better factor American casualties high into the equation. Then senator William Cohen (and future secretary of defense) warned the administration against a heavy ground campaign of attrition, remonstrating, "The notion that we're somehow going to use our land forces to go in and dig the Iraqis out of Kuwait only generates images of young men and women being stacked up like cordwood."[63] There was no political appetite for the campaigns of attrition that had dominated American victories before Vietnam.

Cohen's comments and Dugan's polemic illustrate the tension in the lead-up to the Gulf War between those that preferred campaign plans that leaned heavily on overwhelming force (and a large ground contingent) and those that advocated for surgical and rapid displays of airpower designed to create effects within circles of gravity. Both of these approaches sought to decrease US casualties and ensure public support. For example, one of the thought leaders of the airpower argument, Colonel John Warden, remembered that air campaigns used casualty projections (both friendly and enemy) as an important measure of success. As Warden explained about decision-making behind the air campaign in Iraq, "Reducing adversary casualties and US casualties went hand in hand; we were all horrified by the unnecessary casualties in Vietnam. . . . We also made some assumption that if we had gone in to present in front of Schwarzkopf and Powell and if we had said we can defeat the Iraqis in a short time but kill a million Iraqis, it would have been rejected out of hand. If the plan was going to be accepted, it was the one that would win and would do it with an acceptable loss rate on both sides."[64] While both the overwhelming ground force and airpower philosophies sought to mitigate casualties (especially American losses), they had a fundamentally different assumption about the role of technology in the ability to preserve that public support (and thus win the war).[65] This divergence led to slightly different beliefs about the use of unmanned systems in future conflict.

For those advocating for a large ground force, the focus was on creating overmatch via maneuverability and overwhelming advantages in firepower. By using technology in conjunction with a large ground force campaign, the United States could create decisive battlefield victory, all while protecting its forces (key to strategic success). As Bob Scales explained about the Gulf War campaign strategy, "The Army has a moral obligation to the American people to lessen the cost of the battle in American battle. To honor such an obligation, there can be no such thing as a fair fight. . . . [T]he object of future wars, therefore, will be to collapse an enemy by maneuvering an overwhelming joint force against him so that his will to resist is broken and close-in killing becomes a coup de grace rather than a bloody battle of attrition."[66] Technology in this instance creates overmatch and agility—all by providing a persistent "picture of the battlefield."[67] Unmanned systems, tied to ground units, help create that persistent picture of the battlefield, while missiles help create the overmatch required to dominate the enemy. Both increase the distance from the warfighter to the fight, making unmanned systems an enabler of overwhelming first-strike capabilities partnered with ground units equipped with unmanned aerial vehicles to provide unit-level force protection.

This was primarily a ground-based perspective on the Gulf War, but there was a sometimes competing and other times supporting theory of victory driven by airpower. In this alternative theory of victory, technology enabled not agile overmatch of ground forces but instead effect-based operations to create quick and low-cost victory. In this conception of warfare, an information-age update on strategic bombing, precise munitions launched primarily by aircraft would bring a military and regime to its knees, leaving the Army to sweep in and clean up the surrendering masses.[68] An airpower-driven campaign ensured public support by winning wars primarily remotely. Led by John Warden and an entrepreneurial Air Staff, the airpower campaign was presented as a direct alternative to Vietnam-style air and ground attrition strategies. Even the name was devised to invoke and then repudiate Vietnam. As Warden's former assistant recounted, "The last thing they wanted to do was to repeat the Rolling Thunder strategy. Warden responded: 'That's exactly it: it's not Rolling Thunder—its Instant Thunder!' . . . His plan would be called 'Iraqi Air Campaign Instant Thunder.'"[69] Schwarzkopf loved it and, despite internal Air Force dissension about Warden's circles of gravity theory of airpower, decision-makers were enthralled with a proposal that offered a technological theory of quick victory with minimal American and Iraqi casualties. The underlying belief that technology had reinvented the American way of war was central to Warden's plan. As he explained, the "American belief that technology can solve a lot of things really led over the course of time to American wars that were less bloody than those fought by folks who were not as technologically advanced. . . . Technology has driven to shorter wars that are less casualty intensive."[70]

Both the air and ground approaches featured similar unmanned systems—cruise missiles, precision-guided munitions, and to a much lesser extent (especially for the air proposal) unmanned aerial vehicles. However, the approaches imagined different purposes for these systems. In the ground-focused proposal, unmanned systems primarily provided reconnaissance and support to the ground force with long-range strike via either cruise missiles or manned aircraft–delivered precision-guided munitions to soften the adversary's defenses. In contrast, Warden's plan focused on technology to create overwhelming victory before US forces advanced on the ground. Warden's plan, which was focused more on the target than the means, was largely ambivalent about the platform—manned or unmanned—that would conduct the offensive campaign envisaged by his staff.[71] Missiles could be substituted in some cases for manned aircraft, but in general, the theory relied on long-range munitions and stealth to bring the risk to aircrew to almost zero while the overwhelming success of these strike would necessarily mitigate the risk to friendly ground troops by destroying the enemy's will before even advancing on the ground.

In the end, the Gulf War campaign was more of a hybrid between Warden's heavy strategic airpower focus and an agile but large Army invasion. In practice, unmanned systems skewed toward, on one hand, cruise missiles and precision-guided bombs conducting strategic attacks and, on the other hand, unmanned aerial reconnaissance supporting the ground force. Cruise missiles were particularly key to the early salvos of the war and could have played an even larger role if the United States had not had so few available (and at high cost) in the arsenal.[72] As Schwarzkopf recounted, he had to hold back cruise missile strikes after Powell called from Washington, DC to remind him of the scarcity and expense of the systems. Schwarzkopf (who still saw the utility of these systems to mitigate casualties) subsequently ordered his staff, "Let's not fire any more cruise missiles unless that's the only way to get the job done or you know, obviously if we're going to put a lot of people's lives at risk doing it some other way, we didn't want to do it that way because, although cruise missiles cost you two million dollars they don't cost you any human life."[73] Despite accounting for a relatively small amount of the strikes, precision-guided munitions like the Paveway series laser-guided bombs, the television-guided Walleye and GBU-15, and early electro-optical/GPS SLAM and Maverick missiles were met with significant fanfare; the New York Times heralded the munitions as "a new era in warfare, in which marksmanship counts more than saturation bombing."[74] The Pioneer unmanned aerial vehicles, originally invested in after the shoot-down of Navy pilots in Lebanon, embedded with ground units in both the Army and Marines and laid the seed for the persistent aerial coverage that became such an important characteristic of unmanned systems post-9/11. As the DoD testified to Congress in 1992, "The war marked the dawn of a new technological era. Precision guided munitions proved immensely effective.

Cruise missiles, antiballistic missile defenses, advanced reconnaissance systems and Stealth aircraft were all used successfully for the first time in major combat. . . . American technology saved Coalition lives and contributed greatly to victory."[75]

The Gulf War demonstrated the utility of unmanned systems that had been struggling for position in budgeting priorities at the end of the Cold War, but it was hardly an unmanned war, nor did it represent a revolution in unmanned technologies. It was, however, a pivotal moment for the belief in casualty aversion and campaigns that could protect US forces while winning operations. The lessons from Vietnam had been codified, tested, and had demonstrated their (extraordinary) success. Thinking back to the lessons he took from Vietnam into the Gulf War, Schwarzkopf explained, "It was a nightmare that the American public had withdrawn its support: our troops in World War I and World War II had never had to doubt for one minute that the people on the home front were fully behind them. We in the military hadn't chosen the enemy or written the orders—our elected leaders had. Nevertheless, we were taking much of the blame. We soldiers, sailors, airmen, and Marines were literally the sons and daughters of America, and to lose public support was akin to being rejected by our own parents."[76] The prioritization of American political support for the Gulf War was therefore important to both the war's strategy and how the commanders sought to prepare the American public and decision-makers for potential casualties. As Rick Atkinson concluded in his analysis of the Gulf War, the leaders of the US military "had stayed the course after Vietnam, vowing to restore honor and competence to the American profession of arms and, most important, to renew the bond between the Republic and its soldiery. This—Safwan, March 3, 1991—was their vindication . . . the war had lasted not six weeks but twenty years."[77]

Belief Evolution: Shrinking Budgets, Strategic Vacuums, and the Rise of Force Protection Doctrine

The swift and relatively bloodless victory in the Gulf War seemed to validate the lessons taken from Vietnam. After the campaign, the Bush administration saw record-setting approval levels and the military was given a hero's welcome as many units returned from the Gulf. Drawing on the success and the continued influence of the chairman of the Joint Chiefs of Staff, Colin Powell undertook a major revision of the National Military Strategy in 1991 and 1992. This strategy—the department's first since the demise of the Soviet Union and the end of the Cold War—recognized a new strategic reality for the United States. The United States was a hegemon and, in this new world, could downsize its military force. The military strategy called for drastic domestic and international retrenchments and doubled down on Weinberger's 1984 doctrine of restraint and the requirement of overwhelming military capabilities to commit forces. As the strategy remonstrated, "Once a decision for military action has been made, half measures and confused objectives exact a severe price in the form of a protracted conflict which can cause needless waste of human lives and material resources, a divided nation at home, and defeat. Therefore, one of the essential elements of our national military strategy is the ability to rapidly assemble the forces needed to win—the concept of applying decisive force to overwhelm our adversaries and thereby terminate conflicts swiftly with minimum loss of life."[78] The beliefs in casualty aversion and the need to maintain public support had transitioned from individual beliefs within a generation of officers to a battle-tested doctrine that created the foundational assumptions of a post–Cold War military.[79]

The strategy was put to test almost immediately. While American troops and equipment remained invested in the Persian Gulf to keep Saddam Hussein at bay, they were also deployed in Africa, committed to a mission in Somalia that started out as humanitarian relief and devolved into the protection of those humanitarian activities. Domestic support was never high for the mission (and there was significant debate within the Pentagon about the use of American troops),[80] but it plummeted in 1993 when a team of American Marines was killed in the streets of Somalia after rebels downed a US Blackhawk. Eighteen Americans died, some in a very public display when Somali rebels paraded American bodies through the streets of Mogadishu.

This graphic loss of American lives in a contingency that was already far from popular with the American public was exacerbated by what many (especially military voices) came to call the CNN effect.[81] The advent of the twenty-four-hour news cycle and embedded television reporters had only strengthened military officers' beliefs post-Vietnam in the media's impact on public casualty aversion and distaste for war. Media strategies were included in doctrine, and professional military education touted media effects as a key consideration in operational campaign success. A 1994 Parameters article, for example, warned military practitioners that:

In the age of CNN, future wars and OOTW will occur in real time for both the American people and their policymakers. That this development can have positive results against an enemy was illustrated by the Gulf War. But the more pernicious results in terms of less favorable events up and down that continuum has a long history, whether it be the dismissal of Churchill from the Asquith government after the operational defeat at Gallipoli, the decision of LBJ not to run for reelection as a result of Tet, or the effects of the tactical loss of US Army Rangers in Somalia on the tenure of former Secretary of Defense Aspin.[82]

New president William Clinton's experience in Somalia—which occurred early in the administration's tenure—made a lasting impression on the next seven years of the Clinton administration. Clinton himself claimed that the battle, and the graphic loss of life in particular, had a haunting effect on him, which framed military decisions over his two terms.[83] He withdrew American forces from Somalia by early March 1994. His first secretary of defense, Les Aspin, took the initial fall and resigned from office in 1993. However, unlike the Reagan and Bush administrations' Weinberger and later Powell doctrines, which argued for a selective use of military force, Clinton did not curb the use of the military. Instead, he increasingly looked for technological substitutions so that he could continue to use the military while mitigating political risk. In Iraq, he chose to enforce the status quo with high-tech/low-risk tactics: fighters patrolling no-fly zones equipped with long-range smart missiles and strict rules of engagement accompanied by long-distance missile salvos shot into Iraq from well beyond the border. In Bosnia and Kosovo, Clinton doubled down on these long-range air attacks and cruise missile campaigns, executing operations that called for very few ground troops and air tactics designed to mitigate pilot risk.[84] As Secretary of Defense William S. Cohen and General Henry H. Shelton, chairman of the Joint Chiefs of Staff, asserted to the Senate Armed Services Committee, "The paramount lesson learned from Operation Allied Force is that the well-being of our people must remain our first priority."[85]

The beliefs motivating the Clinton administration from the top down also found their way onto the battlefield from decisions made from the bottom up. The 1990s saw the belief in public casualty aversion evolve and cascade into a new theory of victory, force protection. Force protection defined mission success by the ability to mitigate casualties. And while keeping losses at a minimum has always been an important part of winning wars, force protection doctrine placed casualty mitigation over other measures of mission effectiveness like territory held or objectives seized. As General Wesley K. Clark, who led one of these force protection campaigns, asserted, "It wasn't just the politicians who were pushing the military to avoid casualties. We were feeling the impact of deeply rooted organizational forces from within the military itself. . . . 'Voluntary' operations that incurred casualties might not be sustainable. Period. All senior military leaders sensed it."[86] A vignette published in a 1999 study by the Army War College nicely illustrates this strategic inversion of force protection over other measurements of mission success:

On January 25, 1999, a tall, ramrod-straight young combat-arms officer serving in Bosnia with the 1st Armored Division told the about-to-graduate cadets at West Point, "I tell my men every day there is nothing there worth one of them dying for." It was a startling admission to the cadets who were in the midst of a series of classes on the professional military ethic; the lieutenant's admission was utterly contradictory to what they had been studying. Their studies had led them to believe that minimizing casualties was an inherent part of every combat mission but not a mission in and of itself, particularly one which might impede or even preclude success in the unit's mission—in this case, peace operations within the American sector of Bosnia. Queried by a cadet in the audience as to why he communicated this to his men, the lieutenant responded, "Because minimizing, really prohibiting, casualties is the top-priority mission I have been given by my battalion commander."

Sometime later in the presentation the battalion commander gave his perspective. "It's simple," he said. "When I received my written mission from Division, absolutely minimizing casualties was the mission prioritized as first, so I in turn passed it on in my written operations order to my company commanders. This is the mission we have, this is the environment in which we work."

Some months later an article in Army magazine by an Army major made the same point. Arriving for duty in Bosnia, his brigade commander gave the major the following guidance, "If mission and force protection are in conflict, then we don't do the mission."[87]

The force protection beliefs illustrated in this vignette were a consistent theme of writings coming out of students and professors in the officer professional military education system throughout the 1990s. Articles in

Parameters and *Air and Space Power Journal* and theses from across the professional military education system debated the prominence of force protection and the role of public casualty aversion in US military operations.[88] They accused decision-makers of "force protection fetishism" and often questioned the strength of American casualty aversion. This was a generation that had never served in Vietnam. They saw in the Clinton administration's risk mitigation strategies a force overly focused on mitigating personnel risk and political decision-makers unable to make the hard political choice to not use military force at all.

Despite these young officers' critique of force protection, the concept became solidified in US doctrine and tactics throughout the 1990s. For instance, the Army's AirLand Battle doctrine was updated in 1993 and included extensive references to casualties. As Daniel Johnson notes in his analysis of casualty aversion within US military doctrine:

> *This update to AirLand Battle explicitly urged commanders to minimize friendly casualties, in fact, declaring through its keystone document, "The American people expect decisive victory and abhor unnecessary casualties."* ... *It is worth noting that the 1982 edition of FM 100–5 contains the same number of references to casualties as the 1993 version; but in 1982 two-thirds of these references note procedural reporting of casualties or tactics for reacting to nuclear, biological or chemical attack. In 1993, the manual made two such references. The other references emphasized the importance of minimizing casualties, a key lesson extracted from Desert Storm.*[89]

The Army's updated field manual called force protection one of the four elements of combat power and defined it as the ability to conserve the fighting potential of a force so that it can be applied at a decisive place and time.

Technology played a pivotal role in the idea of force protection. Preferred weapons included those that could insulate casualty risk completely: long-range missiles and high-altitude stealth airpower equipped with precision-guided standoff munitions. When it was not possible to completely remove the man from the battlefield, the doctrine relied on defensive technology like armored personnel vehicles to protect personnel. Force protection used technology, where possible, to substitute for the priceless cost of human life. As Krepinevich wrote about the Army, the service believes "'It is better to send a bullet than a man.'"[90] This was especially true for the Air Force, the service that seemed to benefit the most (and have to make the least sacrifices to their core identity—see Chapter 6) from a technology-focused belief in force protection prioritization. In 1996, Air Force Chief of Staff Fogleman called this "the "new American way of war"....Instead of engaging the enemy in what Fogleman called "a bloody slugfest on the ground," US forces could put greater reliance on their advantages in information superiority and precision strike."[91]

The desire to substitute technology for human risk and the rise of force protection doctrine in the 1990s had a profound impact on the trajectory of unmanned systems. First, missiles—whether launched from the air or the sea—became the linchpin of this new American way of war, often used by the Clinton administration as a preferred risk-free way of combating terrorism and other low-threat bad actors. Second, the desire for precision and mitigation of risk put a premium on systems that could provide persistent, long-range surveillance to generate the targeting information required for this new way of war. Cruise missiles were only as smart as the target locations that humans gave them, and so the 1990s military needed more intelligence options to fight the long-range wars it preferred. This need for intelligence occurred at the same time that overhead satellites became more expensive, their scarce resources controlled by civilian prioritization. The American military of the 1990s needed an option that could provide the kind of situational awareness increasingly craved by military leaders to fight high-precision/low-risk wars—but without the risk of losing a pilot.

The shootdown of Captain Scott O'Grady's supposedly invincible stealth F-117 and subsequent five-day search-and-rescue mission changed American military leaders' risk calculus. If a stealth aircraft could be shot down, how could the US military risk nonstealth manned intelligence aircraft like the U-2? Here was the catalyst that introduced an aperture just big enough to usher in an unexpected new unmanned system, a system not built by military requirements but instead on a hunch by a small company named General Atomics. General Atomics' MQ-1 Predator was slow, not survivable, and not autonomous, but it was persistent and, most importantly, unmanned. The Predator drone was first used in Bosnia, where it was applauded by North Atlantic Treaty Organization Commander Admiral Smith for its ability to protect the force: "The inherent value of flying UAVs is the ability to fly in areas where putting manned vehicles would be unacceptable due to risk or operational considerations."[92]

Clinton's political appointees saw an opportunity in the unmanned aerial vehicles and pushed the services to increase research and investment into the systems. John Deutch, initially undersecretary of acquisitions, technology, and logistics and then deputy secretary of defense, was a big believer in the importance of unmanned systems, particularly unmanned aerial vehicles. As Deutch recounted, "When I came to the Pentagon in the first Clinton term, I was Undersecretary of AT&L and I was concerned about the slowness that Army and Navy were taking up unmanned aerial vehicles for surveillance but also for targeting. The current vice chairman of the Joint Chief of Staff pushed hard for joint unmanned office and saw that as being a very important thing and so did I. The Predator was a project that I watched and had a great deal of interest in and thought it would have the kind of capability that would be interesting."[93] After laying the seeds for a joint unmanned office in his first term, when John Deutch arrived as deputy secretary of defense for his second tour in the Pentagon, he saw the political will to invest in the technology that he had long believed in. "In his view, drones were a potentially lifesaving reconnaissance technology that would have been operational by now if the armed services weren't so myopic or the acquisition system such a mire."[94] Deutch stood up a joint organization, the Defense Airborne Reconnaissance Office, to spur all of the services into adopting Predator-like technology. Influenced by the force protection focus in the department during the 1990s, the office recommended that unmanned systems focus on missions that risked manned personnel, decreasing both the risk to pilots and (ultimately) the cost to an ever-shrinking DoD: "Allotting these 'dirty' and dangerous missions to UAVs decreases the risks to manned aircraft and frees pilots to perform missions that require the flexibility of the manned system. UAVs are a viable alternative as the Services wrestle with the many challenges of downsizing the force structure."[95]

While the Army and the Navy may not have been early adopters, the Air Force's new chief of staff was a huge supporter of unmanned systems—based partly on his experience in Vietnam. As Friend writes, "Fogleman had come into office with a background in flying high-risk reconnaissance missions during Vietnam, and he was convinced that UAVs were the future of that capability. As chief of staff, Fogleman established the 11th Reconnaissance Squadron at the end of July in 1995, just as the Predator was showing its potential in Bosnia."[96] This support—from the service that might have at first glance been the most likely to resist unmanned flight—carved out a space for the new systems, leaving them ideally positioned for a new trajectory when 9/11 occurred.

All of this was helped along by Congress. In particular, in 2000 the chairman of the Senate Armed Services Committee, John Warner, pushed for unmanned aerial vehicles and unmanned ground vehicles within the National Defense Authorization Act. He explained his reasoning:

> When you look at the history of casualties, beginning with almost half a million killed in World War II, over 35,000 killed in Korea, and more than 50,000 killed in Vietnam, and zero combat deaths in Kosovo, in my judgment this country will never again permit the armed forces to be engaged in conflicts which inflict the level of casualties we have seen historically. So what do you do? You move toward the unmanned type of military vehicle to carry out missions which are high risk in nature. . . . The driving force is the culture in our country today, which says, "Hey! If our soldiers want to go to war, so be it. But don't let any of them get hurt."[97]

Warner and his staff were keen to inject investment into systems that would insulate military personnel from risk. They saw unmanned systems primarily through the lens of lessons from Vietnam. As Friend explains,

> Warner's SASC staff director, Les Brownlee, was a retired Army colonel, and it was he who first proposed pushing the services to be more aggressive in developing unmanned systems, particularly for conventional attack missions in ways that would replace manned aircraft. In a memo he wrote later to summarize the motivations for the legislation, Brownlee explained, "While unmanned air vehicles had been around for several years, they were used almost exclusively for reconnaissance and intelligence gathering. The intent was to broaden the thinking for exploiting their use." . . . For both men, unmanned platforms could first and foremost remove pilots from harm, with pilots out of the cockpit, legislators wondered if these unmanned aircraft could do things physically that the human body could not withstand.[98]

9/11 and Operation Iraqi Freedom/Operation Enduring Freedom: Casualty Aversion and Force Protection Intertwine and Internalize

The momentum created by Vietnam legacies, beliefs about casualty aversion, and nearly a decade of doctrine geared toward force protection created

a strong path dependency toward unmanned platforms, long-range munitions, and space assets designed to mitigate risk. By 9/11, these beliefs
about casualty aversion and force protection were so deeply intertwined and
ingrained within defense practitioners' belief systems that they were largely
taken as fact. In the decade or so after the fall of the Soviet Union, these
beliefs had led to the dominance of unmanned munitions like long-range
cruise missiles and increasingly precise bombs launched by stealthy manned
platforms to prosecute American conflicts. Meanwhile, unmanned systems
in space grew in importance but shifted from a strategic focus on nuclear to
supporting the targeting intelligence and long-distance command and control required for unmanned platforms and missiles. Mines (both naval and
land), once a part of American wars of attrition in places like Korea and Vietnam, were discouraged by leaders of the era, like Schwarzkopf and Powell,
who spearheaded international movements to curtail their use. Air defense
missiles like the Patriot, the product of AirLand Battle investments made
in the last days of the Cold War and used effectively in Gulf War I, saw little
use in the primarily air campaigns of the 1990s. Meanwhile, unmanned platforms, like the Predator, were still a novelty intelligence resource—a low-risk
alternative to manned intelligence collection as well as a cheaper and more
dynamic alternative to space-based intelligence assets.

September 11th and the subsequent campaigns in Iraq and Afghanistan
significantly shifted the trajectory for unmanned systems and created the
impetus for a noteworthy evolution of beliefs about casualty aversion and
force protection. In particular, after the initial campaigns, it became clear
that—despite the United States' desire for short and overwhelming wars—it
would be difficult to win quickly, decisively, or with a light tech-enabled
footprint in either Iraq or Afghanistan. The two-decades-long conflicts
seemed to debunk the myth of the new American way of war, and any dream
of a Powell or Weinberger doctrine of overwhelming force, clear political
aims, and a swift exit was not going to materialize in Iraq or Afghanistan.
Decision-makers needed to update their beliefs about technology and casualty aversion. If technology couldn't keep the United States from long wars,
could it at least allow the United States to fight these conflicts with limited
bloodshed?

To add to the challenge for this new generation of warfighters, the new
dominant campaign perspective, COIN, required the military not only to
protect its own forces from needless casualties but also to win the hearts
and minds of the population it was trying to conquer. The new US military

needed to retain public support—all while putting its members further and further into risk through community operations and increasingly forward bases. This new COIN strategy substituted swift and overwhelming campaigns for a new theory of persistence, enabled by technology, both on the battlefield for tactical victory and, strategically, in shoring up public support to ensure the nation's staying power within a campaign. In a way, after the failure of Rumsfeld's transformation initiatives to create quick wins in Afghanistan or Iraq, the US military had returned to an updated theory of attrition in which the United States would win not by creating more enemy body bags than friendly but instead by winning a war of the attrition of political will. Who would lose political will first—the insurgents or the United States?

Unmanned systems provided the potential solution to the new problem of persistence and political will. Military leaders, while leery of force protection edicts of the 1990s or the swift technology theories of victory espoused by both military revolutions leaders and many of the airpower advocates of the 1990s, saw in unmanned systems the ability to make war less bloody, more careful, and therefore less susceptible to the whim of domestic and international audiences.[99] This was a new extension of the casualty aversion belief, one in which unmanned systems could decrease not only the risk of friendly lives lost but also the risk of conflict escalation—allowing US decision-makers to use force even in scenarios that might be politically fraught. As Amy Zegart found in her examination of unmanned systems and coercion, "As wars grow longer and less conclusive, armed drones enable states to sustain combat operations, making threats to 'stay the course' more believable. Drones also offer constant stare and precision strike in the same weapon, increasing the certainty of punishment in ways that may change the psychology of adversaries."[100]

Importantly for the trajectory of unmanned systems (and a key difference between unmanned systems envisioned by casualty aversion beliefs versus military revolutions beliefs), this new approach to technology and casualty aversion prioritized control and persistence over speed and automation. Unmanned technologies, like cruise missiles, which were difficult to dynamically retask, or mines, which relied on automation and machine-enabled decision-making, were suboptimal for the new theory of warfare. Even the integration of autonomous unmanned systems for missions like improvised explosive device (IED) disposal or tactical intelligence was questioned by operators: Could these machines build trust with the local population?

Would there be unintended risk from relying on machine decision-making in COIN operations?[101] Could US personnel trust machines to protect friendlies?[102] Unmanned systems designed for COIN still prioritized human control but relied on technology to decrease the physical vulnerabilities of humans in conflict—whether that vulnerability was fatigue, bodily needs, or the risk of casualties. These systems therefore often required a human in the loop and focused on how unmanned systems could create situational awareness for those humans who were at increasingly long ranges from the battlefield.

One type of system, in particular, exemplified the new expanded theory of casualty aversion and the American way of war. Remote-controlled strike-capable unmanned aerial vehicles not only provided persistence in the air but also, by removing humans from the platform, removed the political risk of losing a pilot in unfriendly territory. As Leon Panetta, former head of the Central Intelligence Agency and later secretary of defense, recalled:

> There is no question that in large measure the capabilities we developed with drones in Pakistan were developed because of the nature of the mission that we were involved in there and the fact that Pakistan made it clear that we could not put boots on the ground or fighter planes to conduct missions there. And yet we had this enemy that was there and planning attacks on our country and we couldn't just allow them to continue to do that without going after those that were involved with the attack on 9/11 and attacks in the future. So it came down to frankly the only alternative to be able to go after the enemy that had attacked our country—drones became the weapon of choice to be able to do that.[103]

This was a new twist on unmanned systems strictly for force protection. Instead of focusing on protecting friendly lives on the battlefield, they were also insulating decision-makers from international political ill will, providing a means of persistence in the face of international opprobrium. As Panetta explained, "There wasn't a deliberate debate about we ought to invest in this because it will lead to less casualties—it was viewed as an additional capability for our forces on the ground."[104]

Luckily for the US military and its campaigns post-9/11, the ideal unmanned platform for persistence, precision, and risk mitigation was already operational, with combat sorties in Kosovo and significant use with the intelligence community.[105] Up until 9/11, the Predator had been almost completely used for intelligence. But with the new campaign in Afghanistan, as well as the counterterrorism mission in Pakistan, the military (and its civilian leaders) was interested in arming the platform. The potential to use the new technology to avoid a Vietnam-type defeat in Afghanistan was enticing. As Correll writes, "General John Jumper, who had been working in the CAOC during the Kosovo campaign, got the idea that UAVs could do more than just watch the battlefield. Jumper had flown the same high-risk reconnaissance missions in Vietnam as the now-erstwhile Chief of Staff, General Fogleman."[106] For Jumper, in particular, the ability to avoid downed pilots and lengthy (and risky) search-and-recovery operations was a strong argument for arming the Predators. In Whittle's account of the development of the Predator, the author recounts a pivotal presentation given to General John Jumper shortly before the decision to use and arm unmanned aerial vehicles in which the PowerPoint slide declared, "The Rules of Have changed," and depicted a Serbian soldier on the downed F-117.[107]

Decision-makers chose to arm the Predator with a relatively small guided missile, the Hellfire, opting for a munition from which they could prioritize control and precision over firepower. The Hellfire, originally an antitank helicopter-fired missile, was initially part of a suite of precision-guided munitions spurred by Perry's second offset. Accordingly, the first few iterations of the missile focused on destroying Soviet tank masses, and subsequent modifications focused on firepower and warhead improvements. By the early 2000s its laser guidance had been adapted to a "fire and forget" system, making it ideal for the unmanned Predator. Subsequent modifications of the missile decreased the lethality of the warhead and updated guidance. The Predator and the Hellfire enabled warfare that was precise, limited, and persistent. The Predator, and then its replacement the Reaper, became an integral part of campaigns in Iraq and Afghanistan and a selling point by DoD leaders to Congress as they advocated for continued support in the wars. As Secretary Rumsfeld told Congress, "Unmanned surveillance and attack aircraft like Global Hawk and Predator offered a glimpse of their potential in Afghanistan. The 2003 budget increases the number of unmanned aircraft being procured and accelerates the development of new unmanned combat aerial vehicles capable of striking targets in denied areas without putting pilots at risk."[108]

The remotely piloted aircraft provided two new capabilities to the US military, which became key to COIN. First, these systems could loiter over targets for long periods of time. Equipped with full-motion video and near-real-time data transfer, tactical units, ground commanders, and campaign leaders could watch the battlefield and targets almost constantly. As former chief of staff Norman Schwartz argued, "It is the ability of remote systems to maintain target area surveillance—long enough for their remote operators to carefully and deliberately evaluate potential targets against the criteria that allow them to shoot—that makes this system so discriminating and effective."[109] This ability to increase the time to make decisions about strikes also helped mitigate casualties, thus ensuring both domestic and international public support for the COIN campaigns. Second, these remotely piloted systems kept pilots out of risk and increased the distance between the shooter and danger, thus decreasing the chance for another Somalia Blackhawk-down event. Once again, as former chief of staff Norman Schwartz argued, "In my mind, the remotely piloted aircraft or RPA is the single most significant advance since radar. This goes back centuries to the bow and arrow. The spear, artillery, aircraft—there's been a momentum to extend the distance between combatants from the very earliest days. The difference is that now we have the ability to fly lethal aircraft without the risk of physical danger to the pilot. It's monumental."[110]

Schwartz's comments foreshadow an important point about beliefs in casualty aversion and force protection. COIN advocated for a campaign of hearts and minds, of US forces embedded within communities, and of risky missions putting US troops in front of and interacting with the population. But all of this came with the potential of human loss. Force protection was certainly not gone, nor were beliefs about casualty aversion—they had just been internalized and evolved as they passed to the next generation of combat leaders. The new generation of casualty aversion beliefs was far more sophisticated, both in its beliefs about the ability to completely reduce the risk of casualties to US personnel and in the ability of technology to decrease that risk. Unmanned aerial vehicles substituted for risk by removing manned pilots. However, over time—and as it became increasingly clear there were not significant threats to manned aircraft—the motivation behind the systems became less about substituting for human risk individually and instead about using unmanned sensors to decrease the chance of loss of life across the conflict.

While insurgents focused on using unmanned systems like IEDs and dumb mortars to create risk and violence, the United States prioritized unmanned systems that could decrease violence, without necessarily sacrificing lethality. On the offense, unmanned aerial vehicles, precise and small-effect missiles, and improvements to laser and global positioning system–guided bombs all helped decrease collateral damage and risk to friendlies. They also provided commanders (on and off the battlefield) with the situational awareness required to enforce tight rules of engagement. Investments in space-based communications and navigation allowed for tight centralization and remote control of unmanned systems, while unmanned aerial vehicles and space intelligence assets previously used for strategic intelligence were pivoted to tactical collection. Meanwhile, on defense, the United States invested in unmanned ground vehicles to detonate IEDs and automated countermortar systems like the C-RAM to shoot down incoming mortar salvos. A new organization, the Joint Improvised-Threat Defeat Organization, was even created to link all these technologies together to counteract IEDs and other insurgent asymmetric tactics.[111]

This was a better version of Westmoreland's invisible fence. It was smarter, more integrated, more persistent, and more controlled than Westmoreland's failed unmanned experiment in Vietnam. But it was the same intention. Despite the lessons of Vietnam and the strength of beliefs in casualty aversion that had been internalized in two generations of officers, the United States was in yet another war of attrition. However, the technological advances in those three decades allowed US decision-makers to adapt unmanned systems to attempt to win a war of attrition while maintaining public support. This led to unmanned systems that were expensive, required significant logistical support, and sacrificed firepower for control and precision. The war also infused unmanned systems with budgetary support, hijacking some munitions and platforms from a much more military revolutions–focused trajectory toward one for unmanned systems that prioritized risk mitigation and control.

Conclusion

Two decades after 9/11 and drawn-out campaigns in Iraq and Afghanistan, "force protection" is no longer a buzzword within US defense discussions, nor are there many who believe that wars can be won solely by long-distance strikes. COIN re-emphasized the importance of human presence in wars of insurgency, which created a natural tension between, on the one hand, the rise of remote-controlled unmanned aerial vehicles and, on the other hand, the human-focused approach of COIN. Even as remotely piloted aircraft dominated US unmanned investments in the post-9/11 era, debates about the mediating role of technology on valor and combat pervaded military discussions.[112] Like late nineteenth-century debates about the torpedo or early twentieth-century questions about airpower, many worried that remotely controlled warfare instilled too much confidence in policy makers bent on prosecuting wars of coercion without escalation. Critics blamed remote technologies and a perception of precision warfare for insulating the American people from casualties—allowing decision-makers to adopt a new American way of warfare that could not rely on technology for overwhelming victory but could make war sterile enough to stay out of the headlines.

After multiple decades in Afghanistan, Iraq, and Syria, few people question the ability of the American population to support the troops in a drawn-out campaign. However, despite the change in perspective about the dominance of force protection or the American ability to endure long-drawn-out conflicts, beliefs in the public's casualty aversion and the linkage between personnel loss and loss of public support (and subsequent loss of strategic victory) are still alive and well within the military. The beliefs created by Vietnam about technology and casualty aversion survived the information revolution, two generations of officers, and a two-decade-long military campaign in Afghanistan. It is a resounding example of a belief that has emerged, cascaded, and internalized. In the end, as John Deutch concluded, "The major impact of Vietnam on the development of weapon systems was the fact that they realized if you didn't have public support or force in the field who were treated well and supported that you had a real problem with the conflict you were involved in."[113] It has become so entrenched and intertwined within an American military identity that it is hard to identify in contemporary strategies as a stand-alone idea or even a noted assumption.

The belief in casualty aversion and later force protection is perhaps so successful because it does not come from a particular thought leader or an influence campaign but instead is passed on through generations of experiences created on the battlefield. This is one of the major differences between the belief in technological determinism and beliefs in casualty aversion, which has direct implications for the trajectory of unmanned systems. The belief in technological determinism was created and cultivated with intentionality by a group of policy wonks deep in the bowels of the Pentagon. They aimed to change acquisition strategies by influencing the services and future civilians to invest in unmanned systems that created advantages in speed, sensing, and automation. They developed narratives and created influence campaigns to ultimately change budget decisions made in the Pentagon. They had to invest in outside experts and think tanks to proliferate their beliefs.

In contrast, beliefs about force protection and casualty aversion were passed down organically by generations of officers with shared experiences. They did not need a narrative, nor are these debates necessarily advocated or debated. Further, their believers are warfighters who build campaign plans, adopt technology on the battlefield, and develop tactics. They are the adopters of unmanned systems and as such create the linkage between investments at places like the Defense Advanced Research Projects Agency or the Strategic Capabilities Office and the actual implementation of unmanned systems on the battlefield. Further, because so much of the budget in the last two decades has been tied to overseas contingency funds, unmanned assets to support COIN, such as remotely piloted aircraft, saw huge influxes of investment, while other systems that might have had a head start (like cruise missiles) were deprioritized and stocks of long-range strike munitions dwindled. When service chiefs didn't agree and tried to refocus technological development and acquisition to military revolutions, they were sometimes fired—as was the case with Air Force Chief of Staff Mosely under Gates and Dugars under Cheney (more about the services in the next chapter).

This isn't the only difference between technological determinism and beliefs in casualty aversion. The military revolutions narrative believes that unmanned systems create revolutionary advances in effectiveness by enabling systems of data that decrease the time from sensor to shooter. Unmanned systems, in the military narratives lens, create decision speed. They do not necessarily mitigate risk to the shooter, except in the sense that by taking the first shot the shooter then destroys the threat before it can cause a risk to them. In contrast, beliefs about force protection create physical—not temporal—distance between the shooter and the threat and stress control and persistence over first-strike advantages. Whereas technological determinism advocates for unmanned systems that speed up decisions, casualty aversion beliefs focus on persistence and loiter so that overall risk to the force is mitigated and decisions that decrease the chance of unnecessary loss of life (both friendly and civilian) are optimized.

Sometimes in our story over three decades the two beliefs elided. For example, advocates of network-centric warfare or airpower enthusiasts argued that revolutionary advances in information technology not only made states more likely to win in tactical engagements but also had the secondary effect of creating swift and costless conflicts.[114] Perhaps one of the best illustrations of this eliding between the two ideas was the concept of disengaged combat, which leaned heavily on John Warden's theories of airpower to argue that "in the future an attack on an enemy's information network would gain victory for the United States 'without firing a shot'...Disengaged combat proposed that increasing the distance between Americans and their enemies enabled American troops to gain victory without the risk of casualties."[115] Therefore, for at least some portion of our story—particularly from the Gulf War until the demise of the Rumsfeld transformation initiative—the two beliefs were mutually reinforcing and often difficult to disentangle in their ultimate effect on both technology and unmanned systems. This was often intentional as military revolution advocates, who had lost significant political momentum under the Clinton administration, piggybacked on force protection doctrines to keep support for military revolutions alive.

For Knox and Murray, two historians funded by ONA and responsible for much of the scholarly work on military revolutions, the two beliefs were necessarily linked. As Jeremy Black explains about the historiography of the RMA, "The movement from mass militaries in the West to smaller professional forces made soldiers fewer but more valuable; while civil society as a whole became more reluctant to regard casualties as acceptable. The two developments were linked, albeit different. They each encouraged a desire for an RMA to provide apparently casualty low, if not free, war."[116] Knox and Murray argued that the historical and techno-determinist understanding of warfare represented a culturally American approach to warfare and controlling the uncertainties of conflict. "The obsessions of the technology utopians derive equally from the deeply and quaintly American belief that all human problems have engineering solutions, and from the profoundly un-American...post-Vietnam search for technological silver bullets that will permit US forces to wage war without suffering—or perhaps even inflicting—casualties."[117]

Sometimes that intertwining between the beliefs occurred within the personnel themselves—individuals with both warfighter experience and a relationship with ONA or beliefs in military revolutions. For instance, James Roche, who served in Vietnam, then worked for Andy Marshall, and finally became secretary of the Air Force, made the final decision to arm the Predator. Frank Kendall, one of the early writers about military-technological revolutions and later undersecretary of acquisitions under Deputy Secretary of Defense Work, spoke candidly about the influence of Vietnam on how he saw technology, explaining, "You always want to avoid casualties; expectation for casualties went down dramatically after Gulf War and we've sustained that since then. I'm a Vietnam era person and Vietnam casualties were dozens a week. I still have memories of the Life magazine with all the pictures. . . . Since then we have had an expectation that low casualties are normal; with an all-volunteer force we have an expectation that casualties are low."[118] Even Bill Perry traced his initial experience with technology and unmanned systems to the bottom-up experimentation he did in Vietnam.

In the end, for contemporary investments in unmanned systems, force protection beliefs dominated unmanned trajectories. The exogenous shock of 9/11, paired with a ready-to-deliver remotely controlled unmanned aerial vehicle, made equipment like the Predator a preferred unmanned technology and created a trajectory for systems that optimized persistence and risk mitigation over speed. But that trajectory and the existence of the Predator—even the strength of both military revolution and casualty aversion beliefs—were all only possible because of decisions made by the armed services. And their identities would ultimately determine how these beliefs translated to the unmanned arsenal the services bought.

Notes

1. Jacquelyn Schneider and Julia MacDonald, "U.S. Public Support for Drone Strikes: When Do Americans Prefer Unmanned over Manned Platforms?," Center for a New American Security, September 2016, https://www.cnas.org/publications/reports/u-s-public-support-for-drone-strikes.

2. Julia Macdonald and Jacquelyn Schneider, "Presidential Risk Orientation and Force Employment Decisions: The Case of Unmanned Weaponry," Journal of Conflict Resolution 61, no. 3 (2017): 511–536.

3. On the role of public opinion in foreign policy decision-making see e.g. John H. Aldrich et al., "Foreign Policy and the Electoral Connection," Annual Review of Political Science 9 (2006): 477–502; John H. Aldrich, John L. Sullivan, and Eugene Borgida, "Foreign Affairs and Issue Voting: Do Presidential Candidates 'Waltz before a Blind Audience'?," American Political Science Review 83, no. 1 (1989): 123–141; Christopher Gelpi, Peter Feaver, and Jason Reifler, "Success Matters: Casualty Sensitivity and the War in Iraq," International Security 30, no. 3 (2006): 7–46; Christopher Gelpi, Peter Feaver, and Jason Reifler, "Iraq the Vote: Retrospective and Prospective Foreign Policy Judgements on Candidate Choice and Casualty Tolerance," Political Behavior 29, no. 2 (2007): 151–174; Christopher Gelpi, Peter Feaver, and Jason Reifler, Paying the Human Costs of War: American Public Opinion and Casualties in Military Conflicts (Princeton, NJ: Princeton University Press, 2009); Michael Tomz, "Domestic Audience Costs in International Relations: An Experimental Approach," International Organization 61, no. 4 (2007): 821–840; Adam J. Berinsky, In Time of War: Understanding American Public Opinion from World War II to Iraq (Chicago: University of Chicago Press, 2009).

4. Matthew A. Baum, "Going Private: Public Opinion, Presidential Rhetoric, and the Domestic Politics of Audience Costs in US Foreign Policy Crises," *Journal of Conflict Resolution* 48, no. 5 (2004): 603–631; Matthew A. Baum and Philip B. K. Potter, "The Relationships between Mass Media, Public Opinion, and Foreign Policy: Towards a Theoretical Synthesis," *Annual Review of Political Science* 11 (2008): 39–65; Berinsky, *In Time of War*; Ole R. Holsti, *Public Opinion and American Foreign Policy* (Ann Arbor: University of Michigan Press, 2004); John E. Mueller, *War, Presidents, and Public Opinion* (New York: John Wiley and Sons, 1973); Dan Reiter and Allan C. Stam, *Democracies at War* (Princeton, NJ: Princeton University Press, 2002); Bruce Russett, "Doves, Hawks, and US Public Opinion," *Political Science Quarterly* 105, no. 4 (1990): 515–538.
5. Mueller, *War, Presidents, and Public Opinion*, 62.
6. John Mueller and John Paul Mueller, *Policy and Opinion in the Gulf War* (Chicago: University of Chicago Press, 1994).
7. For example, Gartner and Segura argue that although human costs are an important predictor of wartime opinion, that temporally proximate costs captured as marginal casualty figures are more important than Mueller's log of cumulative national casualties. See Scott Sigmund Gartner and Gary M. Segura, "War, Casualties, and Public Opinion," *Journal of Conflict Resolution* 42, no. 3 (1998): 278–300.
8. Richard C. Eichenberg, "Victory Has Many Friends: US Public Opinion and the Use of Military Force," *International Security* 30, no. 1 (2005): 140–177; Scott Sigmund Gartner, "The Multiple Effects of Casualties on Public Support for War: An Experimental Approach," *American Political Science Review* 102, no. 1 (2008): 95–106; Gartner and Segura, "War, Casualties, and Public Opinion"; Scott Sigmund Gartner, Gary M. Segura, and Bethany A. Barratt, "War Casualties, Policy Positions, and the Fate of Legislators," *Political Research Quarterly* 57, no. 3 (2004): 467–477; Eric Victor Larson, *Casualties and Consensus: The Historical Role of Casualties in Domestic Support for US Military Operations* (Santa Monica, CA: RAND Corporation, 1996).
9. William A. Boettcher III and Michael D. Cobb, "Echoes of Vietnam? Casualty Framing and Public Perceptions of Success and Failure in Iraq," *Journal of Conflict Resolution* 50, no. 6 (2006): 831–854; Erik Voeten and Paul R. Brewer, "Public Opinion, the War in Iraq, and Presidential Accountability," *Journal of Conflict Resolution* 50, no. 6 (2006): 809–830.
10. Richard C. Eichenberg, Richard J. Stoll, and Matthew Lebo, "War President: The Approval Ratings of George W. Bush," *Journal of Conflict Resolution* 50, no. 6 (2006): 783–808; Peter D. Feaver and Christopher Gelpi, *Choosing Your Battles: American Civil-Military Relations and the Use of Force* (Princeton, NJ: Princeton University Press, 2004); Gelpi, Feaver, and Reifler, "Success Matters"; Gelpi, Feaver, and Reifler, *Paying the Human Costs*.
11. Bruce Bueno de Mesquita et al., "Symposium on Replication in International Studies Research," *International Studies Perspectives* 4, no. 1 (2003): 72–107; Gartner, Segura, and Barratt, "War Casualties."
12. See e.g. Bruce Bueno de Mesquita et al., "An Institutional Explanation of the Democratic Peace," *American Political Science Review* 93, no. 4 (1999): 791–807; Bueno de Mesquita et al., "Symposium on Replication in International Studies Research"; Erik Gartzke, "Kant We All Just Get Along? Opportunity, Willingness, and the Origins of the Democratic Peace," *American Journal of Political Science* 42, no. 1 (1998): 1–27, Erik Gartzke, "Preferences and the Democratic Peace," *International Studies Quarterly* 44, no. 2 (2000): 191–212; Reiter and Stam, *Democracies at War*; Bruce Russett, "A More Democratic and Therefore a More Peaceful World," *Journal of New Paradigm Research* 29, no. 4 (1990): 243–263; Randolph M. Siverson, "Democracies and War Participation: In Defense of the Institutional Constraints Argument," *European Journal of International Relations* 1, no. 4 (1995): 481–489.
13. Feaver and Gelpi, *Choosing Your Battles*, 100.
14. Gartner, "The Multiple Effects of Casualties"; Gartner and Segura, "War, Casualties, and Public Opinion"; Feaver and Gelpi, *Choosing Your Battles*; Gelpi, Feaver, and Reifler, "Success Matters"; Gelpi, Feaver, and Reiter, *Paying the Human Costs*; Bruce W. Jentleson, "The Pretty Prudent Public: Post Post-Vietnam American Opinion on the Use of Military Force," *International Studies Quarterly* 36, no. 1 (1992): 49–73; Bruce W. Jentleson and Rebecca L. Britton, "Still Pretty Prudent: Post-Cold War American Opinion on the Use of Military Force," *Journal of Conflict Resolution* 42, no. 4 (1998): 395–417; Steven Kull, I. M. Destler, and Clay Ramsay,

The Foreign Policy Gap: How Policymakers Misread the Public (College Park: Center for International and Security Studies at Maryland, 1997); Steven Kull and Clay Ramsay, "Challenging U.S. Policymakers' Image of an Isolationist Public," *International Studies Perspectives* 1, no. 1 (2000): 105–117; Larson, *Casualties and Consensus*; Benjamin C. Schwarz, *Casualties, Public Opinion, and US Military Intervention: Implications for US Regional Deterrence Strategies* (Santa Monica, CA: RAND Corporation, 1994).

15. Steven Casey, *When Soldiers Fall: How Americans Have Confronted Combat Losses from World War I to Afghanistan* (Oxford: Oxford University Press, 2014).

16. William Westmoreland, *A Soldier Reports* (New York: Doubleday, 1976); Piers Robinson, *The CNN Effect: The Myth of News, Foreign Policy and Intervention* (New York: Routledge, 2002); Warren P. Strobel, "The CNN Effect," *American Journalism Review* 18, no. 4 (1996): 32–38; Jeffrey P. Kimball, "The Stab-in-the-Back Legend and the Vietnam War," *Armed Forces and Society* 14, no. 3 (1988): 433–458.

17. Carter Malkasian, "Toward a Better Understanding of Attrition: The Korean and Vietnam Wars," *Journal of Military History* 68, no. 3 (2004): 911–942.

18. Casey, *When Soldiers Fall.*

19. Westmoreland, *A Soldier Reports.*

20. Thomas Mahnken, *Technology and the American Way of War since 1945* (New York: Columbia University Press, 2008), 99; H. R. McMaster, *Dereliction of Duty* (New York: Harper Collins, 1997).

21. Mahnken, *Technology and the American Way of War since 1945*, 108.

22. William Conrad Gibbons, *The US Government and the Vietnam War: Executive and Legislative Roles and Relationships, Part IV: July 1965–January 1968* (Princeton, NJ: Princeton University Press, 2014); Henry F. Graff, *The Tuesday Cabinet: Deliberation and Decision on Peace and War under Lyndon B. Johnson* (Englewood Cliffs, NJ: Prentice-Hall, 1970), 81–82.

23. Bill Perry in discussion with the authors, May 6, 2020.

24. James Roche in discussion with the authors, April 2020.

25. Henry Kissinger to Richard Nixon, "Impact of American Casualties on Your Vietnam Policy," June 21, 1969, ND8-1, Casualties, Box 26, Nixon Papers.

26. Gentry, "Casualty Management."

27. Survey by Louis Harris and Associates, "Harris 1966 Survey No. 1522," in Harris Collection, Odum Institute, Louis Harris Data Center, University of North Carolina, https://hdl.handle.net/1902.29/H-1522; Survey by Louis Harris and Associates, "Harris 1966 Survey No. 1531," in Harris Collection, Odum Institute, Louis Harris Data Center, University of North Carolina, https://hdl.handle.net/1902.29/H-1531.

28. Stephen P. Randolph, *Powerful and Brutal Weapons: Nixon, Kissinger, and the Easter Offensive* (Cambridge, MA: Harvard University Press, 2009).

29. Andrew F. Krepinevich Jr., *The Army and Vietnam* (Baltimore, MD: Johns Hopkins Press, 1986), 258.

30. John H. Moellering, "The Army Turns Inward," *Military Review* 53, no. 7 (1973): 68; Brigadier General Edward B. Atkeson, "International Crises and the Evolution of Strategy and Forces," *Military Review*, no. 9 (1975): 47–55.

31. Douglas Kinnard, *War Managers* (Annapolis, MD: Naval Institute Press, 1977), 132–133.

32. William M. Hammond, "The Press in Vietnam as Agent of Defeat: A Critical Examination," *Reviews in American History* 17, no. 2 (1989): 312–323.

33. Michael Mandelbaum, "Vietnam: The Television War," *Daedalus* 111, no. 4 (1982): 157–169.

34. Lyndon B. Johnson, "Remarks in Chicago before the National Association of Broadcasters," in *Lyndon B. Johnson, The Public Papers of the President of the United States, 1968* (Washington, DC: US Government Printing Office, 1969), 484.

35. The advent of television news, combined with a campaign of attrition that leaned on technology to create massive firepower, and an unpopular draft certainly affected public support for the war. However, the actual effect of television coverage on support within the public, or even the relationship between public antiwar protests and decision-maker support for attrition campaigns in Vietnam, is still up for debate. Perhaps less covered, but almost as important, was how television coverage of the antiwar movement affected military decision-maker beliefs about support for the military. The antiwar movement against Vietnam brought together large and vocal swathes of the population. This movement was amplified by television

coverage, and the news segments of hippies and protestors burning flags and carrying provocative posters played an emotionally galvanizing role in the American military (both home and abroad). See Hammond, "The Press in Vietnam"; William R. Berkowitz, "The Impact of Anti-Vietnam Demonstrations upon National Public Opinion and Military Indicators," *Social Science Research* 2, no. 1 (1973): 1–14; Adam Garfinkle, *Telltale Hearts: The Origins and Impact of the Vietnam Anti-War Movement* (New York: Macmillan, 1997).

36. Sam Sarkasian, "U.S. National Security Strategy: The Next Decade," in *The U.S. Army in a New Security Era*, ed. Sam Sarkasian and John Allen Williams (Boulder, CO: Lynne Reiner Publishers, 1990).

37. Hamilton Howze, "Vietnam: An Epilogue," *Army*, no. 25 (July 1975): 12–16.

38. Jonathan D. Caverley, "The Myth of Military Myopia: Democracy, Small Wars, and Vietnam," *International Security* 34, no. 3 (2009): 119–157.

39. John L. Romjue, *From Active Defense to Airland Battle: The Development of Army Doctrine, 1973–1982*, (Fort Monroe, VA: US Army Training and Doctrine Command, 1984).

40. Department of the Army, *FM 100-5: Operations* (Washington, DC: Department of the Army, 1976); Adam Joyce, "The Micropolitics of 'The Army You Have': Explaining the Development of U.S. Military Doctrine After Vietnam," *Studies in American Political Development* 26, no. 2 (2012): 180–204.

41. Department of the Army, *FM 100-5*, 2.

42. Jeffrey W. Long, *The Evolution of US Army Doctrine: From Active Defense to AirLand Battle and Beyond* (Fort Leavenworth, KS: US Army Command and General Staff College, 1991).

43. Joyce, "The Micropolitics of 'The Army You Have.'"

44. Daniel Isaac Johnson, "Death and Doctrine: U.S. Army Officers' Perceptions of American Casualty Aversion 1970–1999" (College Station, TX: Texas A&M University, 2010).

45. Joyce, "The Micropolitics of 'The Army You Have.'"

46. Schneider, "The Information Revolution and International Stability."

47. US Department of State, Office of the Historian, Foreign Service Institute, *The Reagan Administration and Lebanon, 1981–1984*, https://history.state.gov/milestones/1981-1988/lebanon; John H. Kelly, "Lebanon: 1982–1984," in *U.S. and Russian Policymaking with Respect to the Use of Force*, ed. Jeremy R. Azrael and Emil A. Payin (Santa Monica, CA: RAND Corporation, 1996), 85–104, https://www.rand.org/pubs/conf_proceedings/CF129/CF-129-chapter6.html.

48. Friend, "Creating Requirements."

49. Richard S. Halloran, "US Buys Israeli Pilotless Planes," *New York Times*, last modified May 24, 1984, https://www.nytimes.com/1984/05/24/world/us-buys-israeli-pilotless-planes.html.

50. George C. Wilson, "Unmanned Gain Backing," *Washington Post*, last modified September 24, 1985, https://www.washingtonpost.com/archive/politics/1985/09/05/unmanned-weapons-gain-backing/00fd9af9-da42-42c6-b174-6f9f50a08e26/.

51. "The Weinberger Doctrine," *Washington Post*, November 30, 1984, https://www.washingtonpost.com/archive/politics/1984/11/30/the-weinberger-doctrine/c7f20ffe-b591-4189-ad05-a704aac1935d/.

52. Krepinevich, *The Army and Vietnam*, 269.

53. Anthony Zinni, "A Conversation with General Anthony Zinni," interview by Bruce R. Kuniholm and Alex Roland, Duke University Living History Program, August 23, 2016, http://livinghistory.sanford.duke.edu/interviews/anthony-zinni/.

54. Marine Corps, "Unmanned Vehicle Unit Changes Name," news release, February 9, 1996.

55. David H. Petraeus, "Lessons of History and Lessons of Vietnam," *Parameters* 16, no. 3 (1986): 43.

56. Richard Halloran, "Officers Shaped by Vietnam Taking Control of the Army," *New York Times*, June 28, 1987, https://www.nytimes.com/1987/06/28/weekinreview/officers-shaped-by-vietman-taking-control-of-the-army.html.

57. Bradley Graham, "Leader of US Troops in Mideast: An Unconventional Warrior," *Washington Post*, March 6, 1998; Frederick Franks, "Interview with Frederick Franks," *PBS Frontline: The Gulf War*, January 9, 1996; Thomas E. Ricks, *Fiasco: The American Military Adventure in Iraq* (London: Penguin UK Books, 2006); Karl W. Eikenberry, "Take No Casualties," *Parameters* 26, no. 2 (1996): 109.

58. Halloran, "Officers Shaped by Vietnam Taking Control of the Army."

59. Bob Woodward, *The Commanders* (New York: Touchstone Books, 1991); Frank Hoffman, "A Second Look at the Powell Doctrine," War on the Rocks, February 2014, https://warontherocks.com/2014/02/a-second-look-at-the-powell-doctrine/; Colin Powell, "U.S. Forces: Challenges Ahead," *Foreign Affairs*, Winter 1992/1993, https://www.cfr.org/world/us-forces-challenges-ahead/p7508; Walter LaFeber, "The Rise and Fall of Colin Powell and the Powell Doctrine," *Political Science Quarterly* 124, no. 1 (2009): 71–95.

60. Colin Powell, *My American Journey* (New York: Ballantine Books, 1996), 143.

61. Woodward, *The Commanders*, 279.

62. Ibid., 291.

63. Ibid., 339.

64. John Warden in discussion with the authors, April 2020.

65. Luke Middup, "The Impact of Vietnam on U.S. Strategy in the First Gulf War," *Comparative Strategy* 29, no. 5 (2010): 389–404; Luke Middup, *The Powell Doctrine and US Foreign Policy* (New York: Routledge, 2016).

66. Bob Scales, *Certain Victory: The US Army in the Gulf War* (Washington, DC: Potomac Books, 1997), 367.

67. Ibid., 370.

68. Thomas A. Keaney and Eliot A. Cohen, *Gulf War Air Power Survey: Weapons, Tactics, and Training and Space Operations*, vol. 4 (Washington, DC: Office of the Secretary of the Air Force, 1993); John A. Warden, *The Air Campaign: Planning for Combat* (Darby, PA: DIANE Publishing Co., 1998); John Andreas Olsen, *John Warden and the Renaissance of American Air Power* (Washington, DC: Potomac Books Inc., 1997).

69. Olsen, *John Warden and the Renaissance of American Air Power*, 148.

70. John Warden in discussion with the authors, April 2020.

71. Warden largely confirmed the ambivalence of manned versus unmanned platforms in his Gulf War strategy in a discussion with the authors, April 2020.

72. To understand why the United States had so few cruise missiles in its arsenal when the Gulf War kicked off, see Chapter 1 and Chapter 5 for how service identity stymied cruise missile investments for decades.

73. "Oral History: Norman Schwarzkopf," *Frontline*, https://www.pbs.org/wgbh/pages/frontline/gulf/oral/schwarzkopf/1.html.

74. Malcolm Browne, "Invention That Shaped the Gulf War: The Laser-Guided Bomb," *New York Times*, February 26, 1991.

75. "Conduct of the Persian Gulf Conflict: An Interim Report to Congress," April 24, 1992, 1–2, https://apps.dtic.mil/sti/pdfs/ADA249445.pdf.

76. H. Norman Schwarzkopf with Peter Petre, *The Autobiography: It Doesn't Take a Hero* (London: Bantam Books, 1993), 208.

77. Rick Atkinson, *Crusade: The Untold Story of the Gulf War* (New York: Houghton Mifflin Company, 1993), 2.

78. Joint Chiefs of Staff, *National Military Strategy of the United States*, 1992, https://history.defense.gov/Portals/70/Documents/nms/nms1992.pdf?ver=2014-06-25-123420-723.

79. Charles A. Stevenson, "The Evolving Clinton Doctrine on the Use of Force," *Armed Forces and Society* 22, no. 4 (1996): 511–535.

80. Sebastian Kaempf, "US Warfare in Somalia and the Trade-Off between Casualty-Aversion and Civilian Protection," *Small Wars and Insurgencies* 23, no. 3 (2012): 388–413; Louis J. Klarevas, "Trends: The United States Peace Operation in Somalia," *Public Opinion Quarterly* 64, no. 4 (2000): 523–540.

81. Robinson, *The CNN Effect*; Steven Livingston and Todd Eachus, "Humanitarian Crises and US Foreign Policy: Somalia and the CNN Effect Reconsidered," *Political Communication* 12, no. 4 (1995): 413–429; Jonathan Mermin, "Television News and American Intervention in Somalia: The Myth of a Media-Driven Foreign Policy," *Political Science Quarterly* 112, no. 3 (1997): 385–403; John Sims, *Shackled by Perceptions: America's Desire for Bloodless Interventions* (Montgomery, AL: Maxwell AFB, School of Advanced Air Power Studies, 1997); Kevin S. Woods, *Limiting Casualties: Imperative or Constraint?* (Fort Leavenworth, KS: Army Command and General Staff College, School of Advanced Military Studies, 1997).

82. Jablonsky, "US Military Doctrine and the Revolution in Military Affairs," 26.

83. William J. Clinton, *My Life* (New York: Alfred A. Knopf, 2004), 552.

84. Perry D. Rearick, "Force Protection and Mission Accomplishment in Bosnia and Herzegovina" (master's thesis, U.S. Army Command and General Staff College, 2001); Wesley K. Clark, *Waging Modern War: Bosnia, Kosovo, and the Future of Conflict* (New York: Public Affairs, 2002), 437–438.

85. John Correll, "Casualties," *Air Force Magazine*, June 2003, 52.

86. Clark, *Waging Modern War*, 437–438.

87. Don M. Snider, John A. Nagl, and Tony Pfaff, *Army Professionalism, the Military Ethic, and Officership in the 21st Century* (Carlisle, PA: Army War College, Strategic Studies Institute, 1999).

88. Charles Moskos et al., "Casualties and the Will to Win," *Parameters* 26, no. 4 (Winter 1996–1997): 136–139; Eric Ash, "Casualty-Aversion Doctrine?," *Aerospace Power Journal* 14, no. 2 (Summer 2000): 2–3; Rearick, "Force Protection and Mission Accomplishment"; Jeffrey Record, "Force-Protection Fetishism: Sources, Consequences, and Solutions," *Air Space Power Journal* 14, no. 2 (Summer 2000); Laura L. Miller and Charles Moskos, "Humanitarians or Warriors?: Race, Gender, and Combat Status in Operation Restore Hope," *Armed Forces and Society* 21, no. 4 (1995): 615–637; Charles K. Hyde, *Casualty Aversion: Implications for Policy Makers and Senior Military Officers* (Newport, RI: Naval War College, 2000); Woods, *Limiting Casualties*; Michael W. Alvis, *Understanding the Role of Casualties in US Peace Operations* (Arlington, VA: Institute of Land Warfare, Association of the United States Army, 1999); Kenneth Campbell, "Once Burned, Twice Cautious: Explaining the Weinberger-Powell Doctrine," *Armed Forces and Society* 24, no. 3, (1998): 357–374; Harvey Sapolsky and Jeremy Shapiro, "Casualties, Technology, and America's Future Wars," *Parameters* 26, no. 2 (1996): 119–127; Kent Johnson, "Political-Military Engagement Policy: Casualty Avoidance and the American Public," *Aerospace Power Journal* 15, no. 1 (Spring 2001): 98–105; Eikenberry, "Take No Casualties," 109; Richard A. Lacquement, "The Casualty Aversion Myth," *Naval War College Review* 57, no. 1 (2004): 38–58; Cori Dauber, "Image as Argument: The Impact of Mogadishu on U.S. Military Intervention," *Armed Forces and Society* 27, no. 2 (2001): 205–229; Troy E. Devine, *The Influence of America's Casualty Sensitivity on Military Strategy and Doctrine* (Maxwell AFB, AL: School of Advanced Airpower Studies Air University, 1997).

89. Johnson, "Death and Doctrine," 74; Michael McCormick, "The New *FM 100-5*: A Return to Operational Art," *Military Review* 78, no. 5 (1997), 3; Department of the Army, *FM 100-5*, 1–5, 2–1, 2–16, 8–4, 9–4.

90. Krepinevich, *The Army and Vietnam*, 6.

91. John Correll, "Casualties," *Air Force Magazine* 86, no. 6 (2003), 49.

92. Whittle, *Predator: The Secret Origins of the Drone Revolution*, 108.

93. John Deutch in discussion with the authors, April 2020.

94. Friend, "Creating Requirements," 112.

95. Department of Defense, Defense Airborne Reconnaissance Office, *Unmanned Aerial Vehicles Program Plan*, 1994, 2–1, https://apps.dtic.mil/dtic/tr/fulltext/u2/a291628.pdf.

96. Friend, "Creating Requirements."

97. George C. Wilson, "Senate Chairman Pushes Unmanned Warfare," Government Executive, March 6, 2000, https://www.govexec.com/federal-news/2000/03/senate-chairman-pushes-unmanned-warfare/1977/.

98. Friend, "Creating Requirements."

99. James Mattis would famously outlaw the use of "effects-based observation" when he was commander of the Joint Forces Command. John T. Correll, "The Assault on EBO," *Air Force Magazine*, January 2013, https://www.airforcemag.com/PDF/MagazineArchive/Documents/2013/January%202013/0113EBO.pdf.

100. Amy Zegart, "Cheap Fights, Credible Threats: The Future of Armed Drones and Coercion," *Journal of Strategic Studies* 43, no. 1 (2020): 9.

101. Samuel Deputy, "Counterinsurgency and Robots: Will the Means Undermine the Ends?" (thesis, Naval War College, 2009).

102. Macdonald and Schneider, "Battlefield Responses to New Technologies."

103. Leon Panetta in discussion with the authors, April 2020.

104. Ibid.

105. Ibid.

106. Correll, "Casualties."

107. Whittle, *Predator: The Secret Origins of the Drone Revolution*, 108.
108. Department of Defense, *Annual Report to the President and the Congress by Donald Rumsfeld* (Washington, DC: US Department of Defense, 2002), 79.
109. General (ret.) Norty Schwartz and Suzie Schwartz with Ron Levinson, *Journey: Memoirs of an Air Force Chief of Staff* (New York: Skyhorse Publishing, 2018), 321.
110. Ibid., 295.
111. Gentry, "Casualty Management," 246.
112. Cora Sol Goldstein, "Drones, Honor, and War," *Military Review* 70 (2015): 70–76; David Blair, "Ten Thousand Feet and Ten Thousand Miles Reconciling Our Air Force Culture to Remotely Piloted Aircraft and the New Nature of Aerial Combat," *Air and Space Power Journal* 26, no. 4 (May–June 2012): 61–69.
113. John Deutch in discussion with the authors, April 2020.
114. David S. Alberts, John J. Garstka, and Frederick P. Stein, *Network Centric Warfare: Developing and Leveraging Information Superiority* (Washington, DC: Command Control Research Program, 2000), 186; Rearick, "Force Protection and Mission Accomplishment," 40.
115. Johnson, "Death and Doctrine," 74.
116. Jeremy Black, "The Revolution in Military Affairs: The Historian's Perspective," *Journal of Military and Strategic Studies* 9, no. 2 (Winter 2006/2007): 10.
117. Knox and Murray, eds., *The Dynamics of Military Revolution 1300–2050*, 179.
118. Frank Kendall in discussion with the authors, April 2020.

Chapter 5
Service Identity

If war becomes robotic, it gets to everybody's tribe.

—Frank Kendall[1]

In Chapter 4, we explored how beliefs in public casualty aversion and the subsequent dominance of force protection doctrine created a trajectory for unmanned systems that privileged persistence, control, and risk mitigation, ultimately leading to the expensive, remotely controlled unmanned aerial systems that were so central to counterinsurgency campaigns in Iraq and Afghanistan. We also discussed the relationship between force protection doctrine and military revolution beliefs, showing how these beliefs—while presenting very different visions for unmanned systems—often piggybacked off of one another to survive. These beliefs together ultimately created an ideological push for increased investment in unmanned systems.

However, despite the strength of these two beliefs, this push for investment often faltered and did not translate into actual weapon systems. How could two belief systems, spanning fifty years, with advocates across civilian and military elites have such difficulty translating dominant ideas into investment? The challenge for these belief systems was the context within which they were operating. These beliefs did not exist in a void; instead, they were pushing against ideological beliefs so ingrained and internalized that they are better understood as identities. Identities, unlike beliefs, do not require policy entrepreneurs, influence networks, or exogenous shocks to translate into action. Instead, they represent a group's status quo response to the world. Identity creates incentives for in-group and out-group behaviors and beliefs are more or less likely to be adopted or shunned depending on their alignment with these identities. This is why when policy entrepreneurs like Andy Marshall sought to spread beliefs about military revolutions, he had to appeal to key advocates within identity groups, involving them in commissions, seminars, and wargames. And it is also why beliefs like public casualty aversion caught on fire when they coincided with one of the largest

The Hand Behind Unmanned. Jacquelyn Schneider and Julia Macdonald, Oxford University Press.
© Oxford University Press (2025). DOI: 10.1093/9780190064419.003.0006

identity shifts of one of the biggest identity groups in the US military—the all-volunteer Army post-Vietnam. Armed service identity is especially important to understanding how the contemporary beliefs we illustrated in the previous two chapters influence unmanned investments because the armed services control the vast majority of budget decisions.[2] The power of these service identities lies in stark contrast to the belief entrepreneurs for both force protection and military revolutions, who were more likely to manifest influence within strategy or policy (in the case of military revolutions) or battlefield adaptation and contingency funding (in the case of casualty aversion). The influence of beliefs on top-down strategic pushes for unmanned systems, and bottom-up adaptation and unmanned procurement, can be ignored, manipulated, or co-opted when the services sense they might compete with the platforms and projects core to each service identity.

In this chapter we explore how armed service identity intersects with beliefs and manifests in budgets to create, stall, or hijack unmanned trajectories. We first discuss the identities of the four armed services—the Army, the Air Force, the Navy, and the Marines—as well as (sometimes competing) occupational identities within and across these services.[3] We then trace the impact of identity within these services and specialties to detail how each service and occupational identity affects attitudes toward unmanned systems. Finally, we look at how identities interacted with both military revolution and force protection beliefs impacting how each ultimately influenced unmanned investments. This chapter, therefore, retraces the same story as our previous two, but from a very different vantage point. We now turn to the protagonist in these weapons decisions, the gravitational pull around which all of our other characters dance.

Introduction to Identities: Service and Occupational

What are these service identities that play such an important role in the story of unmanned technologies? The seminal work on service identities comes from Carl Builder's 1989 volume, *The Masks of War*. In the book, Builder identifies five faces of the services' personalities: "1) altars for worship, 2) concerns with self-measurement, 3) preoccupation with toys versus the arts, 4) degrees and extent of intraservice (or branch) distinctions, and 5) insecurities about service legitimacy and relevancy."[4] Builder uses these five faces to build out identities, or masks, of the Army, Navy, and Air Force. He argues that these masks explain which technologies the services invest in and why.

For Builder, the Army's identity comes from "its roots in the citizenry, its long and intimate history of service to the nation, and its utter devotion to country."[5] It measures itself by end strength (or total amount of soldiers) and people—not equipment. This leads to a focus in the Army on art versus toys, and pride in tactics and strategy over technology. For the Army, this identity overrides occupational allegiances, with only one major ideational divide between traditional combat arms (infantry, artillery, and armor) and all other support jobs. Builder argues that "the branches have a brotherhood not evident among these specialties in the other services"[6]—resulting in less competition for resources than might occur in some other services. As one of the oldest and most combat-proven services, Builder finds that the Army is the most secure and therefore most likely to adopt joint philosophies, including joint technologies. Together, these characteristics lead to an Army that is "the nation's obedient and loyal military servant . . . prepared to forge America's citizenry into an expeditionary force to defeat America's enemies overseas."[7]

Builder's Navy derives its identity from tradition and a deep belief in its own institutions, the most important of which is independent command at sea described by Builder as "a unique, godlike responsibility unlike that afforded to commanding officers in other services."[8] The Navy measures its strength by the number of ships. In the past, this meant a prioritization of dreadnaughts, but in the more recent age, the Navy has measured its strength by its number of carrier groups. As Builder points out, "Being one capital ship down is to be a 'quart low' with ominous consequences if not corrected."[9] Despite the Navy's focus on ships, Builder asserts the Navy is far less wedded to new technologies than the Air Force (though more so than the Army). Instead, Navy identity and their technological allegiances are tied closely to their intraservice affinities. As Builder explains, "At the pinnacle of this structure, since World War II, has been carrier-based fighter aviation. At (or very near) the bottom is mine warfare. Submarine and surface warfare specialties, in that order, lie in between."[10] The wide range of occupational identities underneath the broader Navy identity allows the service to confidently co-opt new technologies without sacrificing the Navy identity. For the Navy, the bigger pressure on technological adoption is not so much competition with other services as competition between intraservice identities, for example, submarine officers versus carrier aviation. For example, Builder notes that "the threat posed by the airplane was ultimately co-opted by transferring the capital ship mantle from the battleship to the carrier. . . . [T]he ballistic missile was adapted to the submarine, but it has never been close to the heart of the Navy."[11] Cumulatively, these characteristics create a Navy that is focused on "preserving and wielding sea power as the most important and flexible kind of military power for America as a maritime nation. The means to those ends are the institution and its traditions."[12]

Builder doesn't identify the Marines as their own service, but the Marines—officially a subset of the Navy—have one of the strongest and most cohesive identities within the armed services. They are, therefore, an important part of this discussion, and their relationship with the Navy often dictates their identity. The Marines are, by law, a naval service. Their budget comes from the US Navy and appropriations often detail what is considered "blue" Navy money for the traditional sea-based Navy and "green" Navy money for the often land-based Marine force.[13] The Marines are therefore always naval. However, the service has worked hard to create a separate and distinct identity from their naval brethren. For instance, while they both commission officers out of the Naval Academy in Annapolis, naval officers adopt naval ranks (captains, ensigns, admirals) and Marine officers adopt the Army–Air Force ranks of lieutenant, captain, major, and so forth.

The mission of the Marines has changed over time, which suggests that perhaps part of the core identity for Marines is their flexibility as a service to adapt to the needs of naval campaigns.[14] They quickly adopt and change their doctrine based on advances in warfighting, more concerned with showing their utility and relevance over time than their identity within one warfighting competency.[15] This makes them extremely flexible with technological adoption, not beholden to particular platforms or technologies for their core identity. This all leads to a Marine Corps that a 2020 RAND study identifies as having "several competitive goals. First, it strives to demonstrate continued relevance by leveraging forward presence and furthering the idea of the Marine Corps as America's crisis response force. Next, it has the goal of preserving its forcible entry mission for contested environments, as well as a goal of preserving operational responsibility and autonomy. Finally, the Marine Corps has as a goal the preservation of its unique culture, which it sees as an inherent advantage."[16]

Of these services, the Air Force is the newest.[17] While theoretically this should lead to a more nascent and pliable identity, the insecurity created by the service's novelty creates incentives for the Air Force to cling even closer to its core identity. For the Air Force, a service built around the airplane, technology (particularly those that expand air and space flight) is core to identity. Builder explains that this creates a somewhat perverse logic in which "if the Air Force fosters technology, then that inexhaustible fountain of technology will ensure an open-ended future for flight; that, in turn, will ensure the future of the Air Force."[18] The Air Force, therefore, measures its success by its ability to develop and field these flight-related technologies. However, unlike the Navy, which measures itself by the number of ships in its inventory, for the Air Force the focus is quality over quantity. It would prefer less aircraft that are more technologically advanced over more aircraft that do not feature new missiles, radars, or airframes. This leads to Air Force assessments of its own identity based on qualities like stealthiness, altitude, speed, maneuverability, or range instead of number of squadrons or airframes. As Builder describes, "The Air Force, conceived by theorists of air power as an independent and decisive instrument of warfare, sees itself as the embodiment of an idea, a concept of warfare, a strategy made possible and sustained by modern technology. The bond is not an institution, but the love of flying machines and flight."[19]

Branch or occupational identities also play a role in forming service identities and can sometimes compete with service identities for dominance. For the Navy, branch identities cleave closely to identity-defining technologies: for example, the carrier, the submarine, or the destroyer. Branch identities within the Army (e.g. infantry versus artillery) can influence the Army's theory of victory and sway technological preferences. Sometimes occupational identities—pilots, surface warfare officers, infantry, and so forth—can extend across services and compete for influence not only within services but also among services. For example, joint terminal attack controllers, who call in aircraft for ground support, exist within occupations across the services (Army special forces, Air Force and Marine combat controllers, Navy Seals) but exert influence in the identity of the military as a whole. Occupational and branch identities are therefore important inputs to weapons and budget choices and, for unmanned systems, play an important role in both pushing and stalling unmanned investment.

Builder's analysis was published just a few years before the Goldwater-Nichols Act, which significantly realigned budgetary and advisory power within the DoD.[20] Most critically for the role of identity in overall acquisition choices, Goldwater-Nichols attempted to decrease the power of the service chiefs and the role of service identity in weapons choices. However, in practice, it created new competitions over identity and resources, particularly between combatant commands (CCMDs) and the services. As a 2019 RAND analysis of service identity post–Goldwater-Nichols summarizes, "Services remain the most powerful organizational actors in national defense. However, their relative edge over the Office of the Secretary of Defense (OSD), CCMDs, and the Joint Staff has decreased, leading to a more complex field of competition."[21]

For unmanned systems, the post–Goldwater-Nichols military introduced the combatant command as a new actor that competed with the identity of the services to prioritize different unmanned systems. CCMDs shape both the beliefs that are adopted and the ultimate trajectory for unmanned systems by competing directly with service identities. While the CCMDs do not expressly develop new technology or make procurement decisions, they identify and recommend needs for the service chiefs. When there is a disconnect between the immediate needs of these CCMDs and the priorities of the services, OSD or Congress can attempt to intervene, potentially even by firing the service chief. For example, Secretary of Defense Robert Gates removed an Air Force chief of staff (after a nuclear debacle) and appointed a successor far more supportive of Gates's remotely piloted aircraft priorities. Additionally, secretary of defense Rumsfeld circumvented what he saw as antitransformation senior officers in the Army, opting to replace Shinseki (with whom he had a difficult relationship) with General Peter Schoomaker, a reservist with a special operations background.[22]

Congress is the primary civilian arbitrator in this battle of service, occupational, and combatant identities through their oversight of the National Defense Authorization Act. And, while rare, the White House can step in to advocate for particular budget items (e.g. former deputy secretary of defense Bob Work got line-item support from the White House for technologies that supported his third offset initiative). Together, Congress and to a lesser extent the executive can arbitrate the strength of service identities on acquisition outcomes. However, as RAND concludes in their study of service identities in the thirty years post-Builder, "Service institutions may lose bureaucratic battles in the present, but they are the most future-focused entities in the national security enterprise. Because they alone are responsible for raising the forces and building the capabilities to conduct war, services also have the lion's share of the national security budget. While the political appointees at the helm of many parts of OSD will rotate through their positions in a matter of years, services have remarkable continuity of purpose across personnel rotations."[23]

The rest of this chapter looks more closely at service and occupational identities to explain how they have shaped the trajectory of unmanned systems. We also discuss how these identities interacted with the beliefs presented in the previous chapters. This allows us to finally explain the dramatic shift in contemporary unmanned systems: why the Air Force— a service dominated by pilots—became the leader in unmanned aircraft; why the Navy fell behind the Air Force in both unmanned platform and munition investments; and why the Army, despite its embrace of casualty aversion beliefs, still focused on manned platforms and ignored long-range strike munitions.

Army Identity and Unmanned Systems

In general, the Army's approach to unmanned systems is best described as ambivalent; it prefers (but does not push for) unmanned systems that enable dull, dangerous, and dirty missions and pushes back on unmanned weapons with little human control. The Army's focus is on technology that enables primarily human-driven land combat—the proverbial "boots on the ground." It does not seek to substitute technology for manpower, and historically, it has viewed automated or automatic unmanned munitions—like land mines and even rockets—as less honorable and therefore a technology of last resort, used primarily for defense. Their focus on these defensive unmanned missions, air defense missiles for example, is not always core to the identity and has significantly fluctuated in Army priorities as the Army identity reckoned with offensive and defensive campaign preferences.

For the Army, this means they may develop unmanned systems that help mitigate risk to friendly forces, but they do not take the warfighter (the infantryman, the airborne cavalry pilot, the artillery officer) away from the battlefield. Instead, they support core ground warfighting units with unmanned intelligence, communications relays, and ordnance disposal. The focus on unmanned systems as enablers, versus core strike capabilities, means that these systems are not direct substitutions for manned platforms like tanks or ground vehicles. In general, these unmanned enablers (like unmanned aerial vehicles [UAVs]) complement but do not directly compete against existing weapon systems. This means that there is little organized resistance from occupational identities that would see these unmanned systems as a direct threat to their career field. However, because these missions are not core to the Army identity, the Army is also willing to sacrifice them to other services (like the Air Force) when there is an interservice contest for control.

The Army's ambivalence toward unmanned systems means that the pattern of proliferation of unmanned technologies has been influenced more by the Air Force and Navy. Unmanned systems preferred by the Army—those meant to enable ground campaigns—were generally underresourced compared to other unmanned systems that the Air Force and the Navy preferred. This pattern began to emerge after the creation of the Air Force in 1947 as the three services vied for jurisdiction over missiles. Coming out of World War II, there were forty-seven guided missile projects in development. However, a major slash in postwar spending meant that number needed to be significantly cut down and interservice competition dominated the ultimate shape of that winnowing, creating path dependencies for guided munition. throughout the Cold War.[24]

The Army had early successes in missile development. Its Redstone Arsenal created the first missile used in an American nuclear test and the Army's Jupiter intermediate-range ballistic missiles outperformed the Air Force competitor, the Thor. However, when the Air Force launched an effort to control ballistic missiles (including a remarkable public campaign in the *New York Times* maligning the Army's missile program), the Army eventually ceded most of these missions and capabilities to the Air Force. The Jupiter intermediate-range ballistic missiles were ultimately fielded by the Air Force and decommissioned shortly after.[25] The one long-range nuclear-capable missile the Army retained control of, the Pershing (I and II), survived the Air Force onslaught only because they allowed the Air Force to axe the Mace cruise missile (which the Air Force hated even more than the Army-controlled Pershings). When the Pershings were retired in 1991, the Army also retired its interest in the unmanned mission of long-range strike— only recently (and after the Air Force and Navy left the Army with little to do in the shift to the Indo-Pacific) discussing long-range strike missiles again. These systems never occupied a major part of the Army identity. It was never that the Army was opposed to missiles and therefore impeded investments in the systems, but more that the systems didn't have a champion within the Army that would make it core to the Army ground identity.

This pattern also helps explain the Army's investment in missiles for air defense. Like long-range strike missiles, the Army made a first foray in air defense missiles, starting during World War II, spurred on by the Korean War, and fielding the first versions of the Nike air defense missile in 1953. However, the Air Force once again put up a spirited campaign, arguing both that they should control long-range surface-to-air missiles and that the Air Force's fast-reaction combat jets made more sense to defend the continental United States than Army-controlled air defense systems. In the first few decades of the Cold War, this relegated Army air defense to short-range systems, first deployed only in Europe and then a few around key cities and sites in the United States. In the half a century after World War II, the Army fielded only a handful of versions of air defense missiles including the Nike, the Hawk/I-Hawk, the Spartan/Sprint systems (which was decommissioned one day after reaching full operational capability), and finally the Patriot. It wasn't until the Air Force and Army combined forces with the Patriot in AirLand Battle that the Army was able to focus on air defense missiles without the distraction of an Air Force intent on stealing the technologies. The Patriot was the only air defense missile designed as part of a larger Army doctrinal strategy, but even it languished in the two decades of counterinsurgency warfare starting after 9/11.

The Army identity didn't have a huge stake in the battles to control missiles. However, where the Army was interested was in using unmanned systems to improve battlefield reconnaissance on the ground. This came to a head in Vietnam. The conflict was proving difficult; Army preference for attrition strategies was under fire—both by the North Vietnamese and by a US domestic population tired of casualties for a war it didn't fully believe in. At the same time, the advent of the microprocessor opened up opportunities for the Army to invest in unmanned systems that might substitute for manpower and decrease casualties. Aspirations of perimeters of unmanned ground sensors, augmented by unmanned aerial reconnaissance, were supposed to decrease risk to GIs and provide a politically acceptable technological solution for antiwar campaigns galvanized by Vietnam's most bloody ground tactics.[26] Ultimately the technology wasn't ready for the kind of unmanned battlefield support the Army wanted in Vietnam. However, it was a pivotal shift for the Army identity, which was moving to the new All-Volunteer Force. The burgeoning information technology revolution made automation and computing more feasible, and the new Army believed that unmanned systems could insulate its newly professionalized combat forces from the bloody wars of attrition that had sunk the conscription force in Vietnam.

During this important period for the Army's identity, two doctrines dominated Army acquisitions and shaped the trajectory of unmanned systems for the Army: first, Active Defense, which integrated new precision-guided long-range strike missiles and air defense for campaigns prioritizing defense maneuvering, and second, AirLand Battle, which reoriented the Army toward offensive campaigns of technological dominance but relied on many of the same technologies as Active Defense. Both of these documents called for a suite of unmanned systems, ranging from UAVs to new air defense missiles, and finally even unmanned ground vehicles. AirLand Battle had an outsized effect on the trajectory of unmanned systems as it co-opted the Air Force (the Army's primary nemesis in control of unmanned systems), included an integrated acquisition program (the Corps 86 acquisition program), and occurred at a pivotal defense budget glut during the Reagan years.

There were a few notable unmanned programs that came from these two doctrines. First, in 1974, General DePuy (author of the Army doctrine colloquially known as Active Defense) set out the requirements for what would be the Arquila—an Army-owned UAV designed for battlefield surveillance. It survived the transition to the more offensive AirLand Battle as a key component of the new system-of-systems technology program advocated by the doctrine and related acquisition effort. The Arquila program exemplified the ideas behind the new doctrine's focus on quick and decisive long-range strike, enabled by networks of capabilities featuring unmanned sensors and aerial vehicles. But Arquila was not adopted seamlessly among the Army. While most of the Army welcomed the idea of a UAV for reconnaissance, the Army branches couldn't agree on how unmanned reconnaissance fit within their conceptions of warfare. This led to a series of requirements built without concrete understandings about how this new system would work on the battlefield. As Erhard notes, "The UAV would clearly disrupt normal routines in a ground unit, demanding, among other things, more soldiers who did not kill the enemy."[27] Costs for the program inflated quickly, and in 1987, despite general Army support for the idea, Congress cut the system. In the end, "The problem centered on weak program oversight, a direct result of the UAV's lack of fit in the Army's tradition-heavy, rigid branch system."[28]

Arquila may have been a failure. However, AirLand Battle's alignment with the ideas behind the technology-focused military revolutions led to a convergence between the Army service identity and both force protection and military revolution beliefs about the use of unmanned systems.[29] As Peter Mansoor writes about the evolution of Army culture, "After the Gulf War, army leaders embraced the notion of a revolution in military affairs combining digital command and control systems, high-tech intelligence, surveillance, and reconnaissance systems, and guided weapons. They believed that better information and precise weaponry could replace mass on the battlefield."[30] The small-scale, technology-heavy (and airpower-centric) conflicts of the 1990s seemed to affirm a new relationship between the Army and technology. While AirLand Battle's Corps 86 created the foundation for future unmanned investment, the 1990s' Rapid Decisive Operations doubled down on digital technologies. The doctrine envisioned a post–Gulf War US Army with unmanned systems designed to provide surveillance for the deployment of a smaller, lighter, and swifter Army footprint. For example, Force XXI—a mid-1990s Army program to modernize Army procurement of digitally enabled technologies for Rapid Decisive Operations—envisioned UAVs as part of a joint venture to digitize and proliferate sensors to support Army warfighting units.[31] In fact, the Army's Training, Doctrine, and Operations Center concluded that "the tactical unmanned aerial vehicle (TUAV) was the most valuable tactical sensor on the battlefield . . . a powerful contributor to the rapid development of combat power."[32] After-action reporting from integration of the Hunter UAV lauded the use of unmanned systems to support ground operations, asserting that the Hunter "has demonstrated great potential for augmenting organic brigade tactical reconnaissance."[33]

The 1990s showcased the Army's support for unmanned systems in the air. Indeed, the Army was the initial backer of the precursor to the Predator system (before the Air Force's desire to own airborne reconnaissance and Fogelman's advocacy changed the course of the Predator's history).[34] But there was a very Army-specific focus in this investment in unmanned aerial systems. Whereas the Air Force was investing in strategic and long-range operational-level unmanned systems with stateside back-end operators, the Army wanted unmanned aerial platforms organic to the ground unit. Most importantly, in the battle for budgets and operational control, the Army wanted to own the unmanned aerial systems. It knew from its previous competition with the Air Force that, left to its own devices, the Air Force would always use platforms to further their identity of strategic bombing and independence from ground maneuver. Therefore, in the 1990s, when unmanned development lived in the joint Defense Airborne Reconnaissance Office, the Army fought for control of the UAVs and, excited about their use as a tactical brigade reconnaissance platform, tried to make the systems Army assets.

The Army may have been keen to keep the Air Force from co-opting unmanned aerial assets, but it was less enthusiastic about unmanned systems on the ground. In 2000, when Chairman of the Senate Armed Services Committee John Warner pushed the Army to integrate unmanned ground systems into its arsenal, the Army resisted. Then assistant secretary of the Army Paul Hoeper fully supported the rise of UAVs but hedged on the ground systems. "Hoeper stressed that, despite the glittering promise of unmanned combat vehicles, the foot soldier will never become obsolete. He said the air war over Kosovo is being described as 'a defining moment, when, for the first time, a country surrendered to airplanes. The airplanes were terrific, and they won the battle of Kosovo. But you win that battle, and then what? Do you go home? You have to win wars the way you always win them. You have to put soldiers on the ground.'"[35]

Army service identity impeded even their own Army programs. This was especially evident in the Future Integrated Combat System acquisition program that linked both manned components with unmanned systems of sensors, ground vehicles, and aerial vehicles. The system required components to function with each other on an equal playing ground—each element integral to the whole system working. If the Army chose to focus on only one component, the whole thing would fail. Ultimately that's what happened. The Army was not opposed to the UAVs envisioned in the program, but it prioritized a new tank and failed to make similar investments in the information technology and unmanned systems required to make the whole suite of technologies work.

The Future Integrated Combat System highlighted the Army's reluctance to focus on unmanned robotics or networks of unmanned system as opposed to the tank, but it also revealed something important about occupational identities and unmanned systems in the Army.[36] The Army is primarily divided into branches. These branches filter budget requests and provide the primary way in which the Army thinks about investing in an unmanned system. The Army was more likely to invest in UAVs over other unmanned ground vehicles because these systems didn't threaten a core identity or mission. The Army was primarily reliant on the Air Force for tactical intelligence—UAVs represented a new complement and supplement to Army core missions, not a replacement. However, because these technologies didn't fall within the confines of a known or powerful branch mission, they also lacked a budget supporter. The Army may have wanted UAVs, but without a branch to advocate for the systems, the technologies often fell into the cracks. As a RAND postmortem found, "Army branches are used to writing requirements to optimize capabilities within their functional areas. But designing an integrated unit from the ground up necessitates prioritizing unit over individual system performance, and optimization of the brigade is rarely compatible with optimization of every individual component."[37]

The conflicts in Afghanistan and Iraq provided an external push for these systems. The appetite for unmanned aerial systems to support ground combat units was voracious, and the Air Force's control over most of these resources (especially in the early years of these campaigns) was frustrating for the Army. Even as the Army invested in smaller unmanned aerial systems that it embedded within their units, they continually clamored to Congress and OSD for more unmanned air support. Former head of Air Force Intelligence, Surveillance, and Reconnaissance General Deptula remembered, "Bob Work wanted more unmanned capability and rang Gates and Gates asked the Air Force for more orbits of MQ-1/9, not realizing that the US Army already had the capability to do this in their inventory. But because they were organically assigned the companies wouldn't give them up. Simply because of different organizational constructs and inability to enforce joint operations, USAF went and bought more capability for these orbits despite the fact that this existed within the US Army."[38]

But even as the Army clamored for unmanned air support, it was less interested in the use of robotics or unmanned systems on the ground. The Army devoted significant resources in combating the improvised explosive device (IED) threat to manned convoys—developing better armor and counter-IED systems for troop carrier vehicles, requesting more unmanned air intelligence support, and modernizing electronic sensor collection and coordination. However, its investment in unmanned ground vehicles largely stagnated during this period. As former secretary of defense Leon Panetta explained, "I think there was a willingness to research in robotics and other areas but just the nature of the mission conducted by the boots on the ground meant that there was a sense that while it was futuristic to talk about a robotic war, the reality was that the missions in Afghanistan etc. had to be handled by human boots on the ground. It was never something that could replace that."[39]

The Army identity played a large role in the types of unmanned systems that the Army ignored, invested in, and ultimately adopted. It explains why the Army is keen to invest in UAVs but has not staked enough of its identity in any unmanned system to control the technologies when there is significant competition from other services. This helps explain why the Army has been less focused on air defense missiles, long-range strike missiles, and even land mines even as it sought (but didn't always fight for) unmanned technology to support core Army missions. This identity interacted with beliefs about force protection and technological revolutions. The post-Vietnam Army shifted away from attrition and toward offensive campaigns of overmatch—both of which privileged force protection and technology as a substitution for manpower. Meanwhile, the Army's role as the primary creator of military power was threatened by a series of low-intensity insurgencies and technological substitution in the 1990s.[40] Even into the early days of the 2000s, transformation initiatives threatened the Army's identity, promising to draw down ground-based forces and reinvent warfare as long range, high technology, and dominated by the air and space domains.[41] It was only with the resurgence of ground warfare in Iraq and Afghanistan that the Army regained its prominence in military power and was able to focus once again on soldier-led campaigns—investing in (and once again battling) the Air Force for control of unmanned aerial reconnaissance to support ground campaigns.

Air Force

If the Army's drive for unmanned technology is shaped by its identity as a citizen army with boots on the ground, the Air Force's unmanned trajectory is all about air and space technology above the human element in war. While the Air Force's dominant occupational specialty, the fighter pilot, is often credited with stymieing unmanned development, the Air Force as a whole has led investment in unmanned systems in the contemporary US military. This started with large missile inventories in the Cold War and satellites in space and led up to today's conflicts, in which the Air Force owns and operates the most UAVs of all the services. As RAND points out "While a preference for manned aircraft is rooted in the Air Force's legacy as the Army Air Corps before 1947, the service's emphasis on innovation has allowed it to adapt to shifts in technology beyond manned flight, even when embracing such changes has not been easy. Advanced technology is a critical component of what it means to be an airman, regardless of specialty. Indeed, dedication to innovation, or using advanced technologies to address national security problems, has been described as a unifying element of Air Force culture."[42]

Technology and airpower may be the Air Force's core identity, but concern for service autonomy and survival is the Air Force's core insecurity. Because it is the newest force and has historically clashed with both Pentagon senior leaders and the Army about the use of airpower, the Air Force has a keen sense of self-preservation. That desire for survival means that the Air Force will sacrifice its core identity (airpower as a strategic force) to gobble up capabilities and missions that ensure the service's survival. The Air Force knows that if, for example, they don't provide airpower to support ground operations, eventually some senior civilian will give the function to the Army, thus eroding the strength and threatening the survival of the Air Force. This leads the Air Force to invest in technologies that are "multimission"—able to give just enough support to the ground to keep the Army at bay while investing in the kind of exquisite technology that the Air Force wants to buy to enable the strategic targeting, decisive campaigns it wants to fly. Alternatively, it motivates the Air Force to co-opt technologies that could otherwise be owned by either the Army or the Navy. For example, at the onset of the Cold War, the Air Force was focused on jet airpower, but the rising capabilities in missile technology (and the Army and Navy interest in owning and controlling ballistic missiles) convinced the Air Force to repurpose missiles as a strategic element of airpower (thus pushing the major cultural division within the Air Force between strategic and tactical air command).[43]

Because the Air Force is so eager to both survive and buy the technology that it prefers for strategic airpower, it is one of the most savvy services about influencing political budgets. Early on it invested in outside think tanks and policy wonks (like Santa Monica's RAND Institute)[44] to use new statistical models to justify airpower requests. Further, its history features strong identity entrepreneurs like Hap Arnold, Curtis Lemay, and Billy Mitchell, willing to break rules to promote the Air Force's service identity. In fact, the idea of rebel leaders is almost as much a part of the Air Force's identity as its love of the airplane and strategic airpower. Of course, this history is also replete with chiefs of staff and identity entrepreneurs whose careers were cut short by their political campaigning for airpower campaigns and new fighter jets. As the RAND report on service identity argues about the post-Builder Air Force, "Resource competition is a preferred arena for the Air Force. By using strategic analysis, the Air Force is able to build strong arguments for resources. The Air Force is comfortable in the competition for future institutional security, which it engages in by building arguments related to near-peer competitors."[45]

But the push for technology and even strategic airpower does not occur equally across the Air Force. Occupational specialties—especially those tied to specific platforms—play a major role in Air Force identity. Members wear patches, scarves, and undershirts that create rallying symbology for these platform identities (even when Air Force leadership outlaws such displays, these symbols of identity continue to appear). These platform-specific identities create asymmetric incentives for technology investments. In the past the dominant occupational specialty was jet pilots. For example, looking at occupational specialties and technology investments in the 1990s, Carl Builder concluded, "The Air Force has been over-focused on airplanes— on *combat* airplanes, on *manned* combat airplanes, on *fast* manned combat airplanes—to the detriment of many systems and capabilities."[46] Since the 1990s the jet pilot identity has become more nuanced with alliances to specific fighter or bomber jets, all while a new tribe of remotely piloted aircraft pilots and cyber warriors increasingly grew in importance. Former secretary of defense Leon Panetta, who would become a large arbitrator of these clans in his final years, remembered, "As you can imagine in the Air Force it was a hell of a debate because most people like to fly their F15s and their F16s and they want to fly their B1 or B2 bombers and the idea that we would deploy those on an unmanned basis was a lot more controversial for the Air Force."[47]

The Air Force's tension between identity (both service and occupational) and unmanned systems can be traced back to the origin of the service and strategic theories of airpower in the aftermath of World War II. As Builder explains, "The first real challenges to airpower theory, as originally conceived by the aviators, with the airplane as the unique means for fulfillment, came out of World War II in the form of Hitler's 'vengeance' weapons the V-1 cruise missile and the V-2 ballistic missile. . . . [T]he advent of guided missiles—cruise and ballistic—and space satellites posed significant new capabilities to displace the airplane as the principal means to fulfill many of the ends of airpower."[48] As we previously discussed in Chapter 2, ballistic missiles were a direct competitor to manned (and unmanned) airplanes. For the Air Force, this became an existential threat to its identity very quickly. If missiles could create the same kind of strategic effects as the bombing campaigns of World War II—and without the high aircrew loss rate—then they could not only divert resources from the airplane but also threaten the rationale for the new Air Force service completely.

The leaders of the new Air Force had to come up with a rebranding campaign that could preserve the airplane and strategic bombing while also allowing for (and owning) surface-launched missiles with increasingly long ranges and heavy firepower. The solution—at least initially—was to make an argument that long-range missiles, with transcontinental routes that required aerodynamic maneuvering, were actually an extension of airpower. Theodore von Karman (a scientist during World War II and colleague of Hap Arnold) explained how the Air Force made this leap: "We used the term 'pilotless aircraft' to cover all types of missiles, so as to prevent the project from falling into the hands of the Army, from which the Air Force was about to separate."[49] The intercontinental ballistic missile went on to live under the Air Force's Strategic Air Command, home of nuclear-equipped bombers and intercontinental ballistic missiles. Strategic Air Command and its early advocate and leader Curtis Lemay dominated the Air Force's identity, especially in the formative years of the new service.[50] A consummate institution builder, Lemay not only proselytized the strategic use of airpower but also grabbed any and all technology that fit under the guise of long-range reconnaissance, warning, or strike. Building on the legacy of Lemay, over the next few decades, satellites, missiles, and manned bombers all became part and parcel of the Strategic Air Command umbrella. As Timothy Schultz explains, "For advocates of an independent air force, strategic bombardment was the essence of their art. . . . [T]echnological advances in automation, remote control, and cruise missiles revealed that airpower could include unmanned weapons. . . . [T]o help secure independent status, therefore, the air force sought dominion over pilotless weapons."[51]

Vietnam directly confronted the Air Force's perception and use of unmanned systems as primarily strategic extensions or complements to bombing. While the Army, Westmoreland, and political elites pushed hard for unmanned sensors and platforms to perform dull, dangerous, and dirty missions to decrease ground losses, the Air Force's identity was still tied to strategic airpower and the legacy of LeMay. By Vietnam, the Air Force had comfortably co-opted long-range missiles and satellites. The Air Force had also made some forays into long-range unmanned reconnaissance, but the cost and technology required to create these UAVs were always far higher than other unmanned options like satellites. The use of unmanned systems for tactical ground support, a part of experiments during World War II, largely languished leading up to and through the early parts of the Vietnam War.[52] As Mahnken explains in his examination of the use of unmanned systems in Vietnam, "Technology was refracted through the prism of the Air Force's organizational culture throughout the war. Because bomber and increasingly fighter pilots dominated the Air Force, air superiority and strike aircraft received the greatest emphasis."[53]

But Vietnam also shifted the service identity of the Air Force. Pilots increasingly had to fly for the Tactical Air Command to support calls for airpower (both politically and from the battlefield). This decreased the power of strategic airpower within the overall Air Force identity and tilted toward collaboration and support with the Army.[54] Further, after significant losses of manned fighter and bomber aircraft in Vietnam, units in combat were receptive to experimentation with unmanned aerial systems to support reconnaissance, decoy, and suppression of enemy air defenses. The Air Force saw the Army's desire for these systems, and this competition once again led the Air Force to take ownership over UAVs that might otherwise find themselves in the Army's hands.

In the period between Vietnam and the Gulf War, unmanned technology largely languished within the Air Force. While the Air Force increased collaboration with the Army and partnered on its AirLand Battle doctrine and subsequent Corps 86 program, it focused its attention on precision munitions as opposed to unmanned platforms. Meanwhile, Strategic Air Command dominated the Air Force's identity as the nuclear-focused command doubled down on strategic capability to deter a nuclear Soviet Union. Therefore, the post-Vietnam era saw the resurgence of focus on ballistic missiles and munitions over the tactically focused UAVs that had gained traction during Vietnam. As Marc Lavore finds in his evaluation of unmanned proliferation during that period,

The Air Force, thus, enjoyed a monopoly on drone and aerial photo reconnaissance capabilities in the Vietnam War's immediate aftermath. With no other agency poised to challenge the Air Force for the mission, that service's leadership could afford to neglect drones. They consequently deactivated

three of their four drone units between 1973 and 1976. Then, in 1977, the Air Force canceled its two premier drone projects—the BGM-34 C and Compass Call—when they experienced cost overruns and reliability problems. Although the technical challenges these programs faced were real, they were arguably not more so than those encountered by other sophisticated weapons programs. However, in the absence of either a pro-drone community within the Air Force or inter-organizational competition forcing the Air Force to stay the course, the Air Force's leadership seized upon hiccups in both drones' development to cancel them. Finally, then, in 1979 the Air Force disbanded its last drone unit. As a result, without the original incentive structure that had driven drones' development and with no meaningful competition from other organizations, UAV development was allowed to wither on the vine under a brief Air Force monopoly.[55]

The Gulf War and John Warden reinvigorated strategic airpower and its dominant role in Air Force service identity. Warden and his staff (most notably future head of Air Force Intelligence, Surveillance, and Reconnaissance David Deptula) saw the development of precision-guided munitions as the ultimate technological shift toward strategic air campaigns. Leading up to the Gulf War, he advocated air campaigns based on five concentric rings of strategic targets. He promised campaign leaders that effective targeting and stealth aircraft could strike inside the most pivotal leadership rings and effectively cut the head off the snake.[56] The logical extension of this was that strategic use of airpower could create both decisive and relatively bloodless campaigns. Warden's concept of operations seemed to be validated during the Gulf War. And while the enterprising colonel (who had made enemies across the Air Force as he spread his strategy from the Pentagon) didn't survive to propagate his ideas within the force, Deptula became a pivotal entrepreneur for Warden's theory of airpower. His "effects-based operations" dominated strategies of military power for the next decade.

For the Air Force and its unmanned technology, Warden's and later Deptula's strategies codified an identity that tied airpower to strategic success, enabled by precision-guided munitions and networks of sensors. The new strategic airpower identity aligned closely with Marshall's military revolutions, and the seeming dominance of airpower in conflicts like Kosovo further solidified an Air Force identity tied to airpower as a strategic capability. For unmanned systems, this was an impetus for investment in long-range precision and increasingly automated munitions. In the aircraft, major advancements were made in radars and avionic suites, all taking the pilot further away from the battlefield but not necessarily further away from the platform.

Indeed, the occupational specialty of the fighter pilot dominated the 1990s and early 2000s. With the demise of the Strategic Air Command and Tactical Air Command after the Cold War and the stark budget years of the post–Cold War decade, fighter pilots emerged as the dominant identity in the new post–Cold War Air Force. Meanwhile, as political leaders and the Army sought to insulate themselves from what they saw as fickle public support for US personnel losses, the Air Force took on the mantle of force protection. US airplanes, equipped with stealth, top-of-the-line radars, or long-range semiautomated missiles, could operate with very little threat in the low-scale conflicts that the United States opted into during the 1990s. Despite the transformation vision of networks of unmanned systems, the lessons that most of the operational Air Force took from Kosovo was that the future of warfare was dominated by the manned airplane.

But there was something happening in the Pentagon at the same time that the manned airplane was dominating conflict in the 1990s. The Army and the Marines had invested in unmanned aerial systems that they used in the Gulf War to support ground units with persistent aerial reconnaissance. At first, the Air Force wasn't interested. However, the Army started to get serious about these systems within their Force XXI program, and the joint Defense Airborne Reconnaissance Office threatened to consolidate these unmanned systems in the joint (or worse, Army) sphere. Similar to the embrace of missiles after World War II, the Air Force saw a threat to its dominance and needed to make a dramatic move to capture the UAV mission. In 1996, Air Force chief of staff Fogelman declared that "on my watch, the Air Force will embrace UAVs and work to fully exploit their potential."[57] Fogelman made the institutional move to take over the UAV program, grabbing both the Global Hawk and the Predator into the Air Force's control.[58] To overcome within–Air Force resistance to these programs, Fogelman explained to his coterie of Air Force generals that the Marines and Army were hot on the heels of this new technology and threatened to hijack the trajectory of these systems. Faced with sister service competition that threatened the Air Force's core service identity, Air Force generals jumped on the UAV train.[59]

Fogelman was a pivotal character in the Air Force's unmanned journey, in large part because his experiences—first in Vietnam and later with the shoot-down of Captain Scott O'Grady's F-117 in the Serbian conflict—shaped his vision of unmanned systems for the Air Force. In 1994, Fogelman commissioned a group of academics and strategic thinkers to run a Defense Science Board study on the future of the Air Force. Entitled "New World Vistas," its authors argued that "current unmanned aircraft have limited capabilities, serving either as cruise missiles or as relatively expendable reconnaissance probes." Instead of these uses (primarily how the Army viewed unmanned systems), the authors proposed a new Air Force version of unmanned systems, explaining that "New information technologies . . . are likely to soon allow the creation of uninhabited combat aerial vehicles (UCAVs) flown by pilots who never leave the ground." The board recommended that the Air Force's new unmanned systems replace its premier manned platforms, arguing that the physiological advantages of unmanned systems would allow for more maneuverable, faster airframes to enable strikes on high-value targets within minutes.[60]

The Air Force had co-opted unmanned systems from the Army. In doing so, they broadened the Army's focus on UAVs for combat reconnaissance only to include the expanded use of unmanned systems across strategic and strike missions—missions which were more closely akin to core Air Force beliefs about airpower and strategy.[61] The Predator—one of the programs hijacked from the joint Defense Aerospace Reconnaissance Office—was subsequently flown by the Air Force in Kosovo, and for a few years the Air Force could comfortably control unmanned assets without putting significant institutional resources behind them. The Air Force's identity and its relationship with unmanned aerial systems on the margins seemed secure.

But 9/11, two long campaigns in Iraq and Afghanistan, and a new secretary of defense, Robert Gates, threatened the Air Force's identity and its relationship with unmanned systems. First, the long attrition campaigns in Iraq and Afghanistan severely undermined the strategic airpower narrative that had been en vogue since Warden's influence during the Gulf War. This was a problem for Air Force identity, especially as airpower in the campaigns became less and less about striking regime or insurgent centers of gravity and more and more about providing persistent reconnaissance and strike support to US ground forces traversing urban and insurgent territories. The Air Force repurposed its fighter and bomber inventories, equipping them with highly precise bombs with smaller and smaller warheads, eventually adopting smart bombs with no warhead at all. Fighters designed for dog-fighting or stealthy strategic attacks flew reconnaissance circles in the sky, searching for IEDs and waiting to be called for shows of force. Air Force strategists argued for the use of these systems within the newly adopted counterinsurgency doctrine, explaining how Air Force exquisite technology could be repurposed for revolutionary capabilities to win a long attrition campaign of political will.[62]

The Air Force's new narrative and rationale for platforms like the F-22 and F-35 was becoming a hard sell as the Predator (and later Reaper) UAVs not only could loiter for longer than their manned counterparts but also were now armed with the accurate and small warhead Hellfire missile. The Preda-tor and Reaper couldn't survive against any real air threat and (especially in the early years) were vulnerable to transmission lags and interference. However, despite these issues, and blessed with a benign air threat envi-ronment, Army ground commanders increasingly clamored for the systems. The real-time video feed from the platforms' sensors allowed the Army to centralize situational awareness (and command), and the slow platforms were able to follow ground units without circling in high fast-combat air patrols that quickly burned limited fuel stores. Most importantly to the Army, the Predator was not primed for strategic attack or air-to-air engage-ments but instead optimized for battlefield reconnaissance against insurgent forces. In the Predator and Reaper, the Army saw the ground-focused airpower that supported its core service identity of land-focused citizen armies.

The Army made their case at the highest levels, ultimately threatening the service identity of the Air Force by contributing to the demise of a key Air Force identity promoter, chief of staff General Michael Moseley. This threat came from a new secretary of defense, Robert Gates, who took office in 2005. A veteran national security leader, he was not naive about the Air Force's ability to co-opt technologies and narratives to get budgets that supported their favorite aircraft. Gates had tried to push for drones in 1992 as head of the Central Intelligence Agency and butted heads with the Air Force even then, recalling, "The Air Force wasn't interested. People join the Air Force to fly airplanes, and drones had no pilots."[63]

Gates's chief of staff, Michael Moseley, was an F-15 pilot with fighter weapons school experience. Moseley wasn't interested in expanding the Predator inventory but instead was focused on acquiring the fifth-generation manned aircraft the service had been developing for strategic attack against a peer competitor (the F-22s and F-35s). As Norman Schwartz, Moseley's replacement, reminisced, Gates "felt like he was butting heads with a service whose culture seemed to impede his objectives—in this case the rapid expansion of RPAs."[64]

Schwartz, Moseley's replacement, was an outsider to the core service identity of the Air Force. He came from the Special Operations Command and, while a pilot, was more aligned with ground perspectives on combat than the Air Force's usual preoccupation with strategic airpower. Schwartz, in his memoir, identifies this outsider quality as the primary motivation for Gates's appointment of the special operations officer to the top echelon of the Air Force. As Schwartz recounts, "Strolling down the Arnold Corridor of the Pentagon, Gates glanced up at the long row of ornately framed oil portraits lining the wall, and thought about the men who had led his Air Force since its inception in 1947 . . . bomber, bomber, fighter. . . . [E]ach of the eighteen prior chiefs had been either a fighter or bomber pilot. It was time for a fresh perspective."[65] While service identities often shape budgets, OSD always holds a trump card—their ability to hire and fire the top leaders of the services. Gates's decision to replace Moseley, a core Air Force fighter pilot, with Schwartz represented a major effort to change the Air Force service identity and, with it, the trajectory of unmanned systems.

Schwartz was a big believer in unmanned systems designed to support ground conflict. Like many in the Army who had invested in unmanned systems to mitigate risk to ground troops, Schwartz saw the Predator and the Reaper as largely force protection assets. As he recounted when explaining his push for unmanned systems within the Air Force, "Lives were being lost and you can't help but ask yourself how many can be saved by adding just one more CAP, or five more, or ten more."[66] While Schwartz made the push for unmanned units and personnel, he also appointed generals who supported his vision for the systems. General Deptula, previously one of Warden's proteges and the author of the effects-based operations doctrine, had become a true believer and was now leading the Air Force's Intelligence, Surveillance, and Reconnaissance capabilities. While he was a career F-15 pilot, he became one of the strongest proponents of unmanned systems in the Air Force, demonstrating how the Air Force's core desire for survival and love of technology can overcome occupational specialty identities. As he explained, it "doesn't matter what specialty or aircraft you have experience in. Senior decisionmakers will do what is best for the country irrespective of bias."[67]

Twenty years after 9/11 and over a decade since Schwartz took over as head of the Air Force, the Pentagon pivoted back to peer competition and Asia. The third offset reinvigorated the military revolution narrative, while almost two decades of counterinsurgency eroded beliefs about ground combat and long campaigns of political attrition. AirSea Battle vaulted the old Air Force identity and strategic attack back to the forefront of the Pentagon and budgets.

After the investing so heavily in Predators and Reapers—long-range unmanned reconnaissance designed to support ground units—spending on these systems stagnated in the new China-focused budgets. Instead, the Air Force co-opted narratives again, taking the third offset and revolutions in military affairs redux to invest in unmanned technologies that both enabled its premier fighters and replaced them. Counterintuitively, while the fighter pilot remained the dominant occupational identity, the growing influence of civilian policy wonks like former head of Air Force acquisitions Dr. Will Roper created incentives to develop unmanned systems that mimicked or replicated an airplane's function. Roper's focus on automated dogfighting revealed the service's desire to retain airframes—even if unmanned—as the core Air Force weapon. Once again, the quest to preserve service identity in the face of occupational identities shaped the trajectory of Air Force unmanned investment.

In the end, the Air Force's service identity had a bit of a Jekyll and Hyde influence on unmanned trajectory. While the core service identity of technology and service autonomy led to the adoption of unmanned systems in the face of service competition, occupational specialties (both the protection and obsession with replacing core identities) sometimes slowed or changed the direction of unmanned investment. Former secretary of defense Leon Panetta sums up this tension well, explaining that the Air Force has been "the point of the spear for developing these unmanned systems, they perfected the piloting of these systems. The Air Force, combined with intelligence and special forces are really the combination and impetus for unmanned systems."[68] However, Panetta goes on to identify sects within the Air Force as some of the primary inhibitors of unmanned systems.

How did a service filled with pilots become the biggest adopter of unmanned aerial technology? The answer is that the Air Force identity has always been about technology agnostic of the man. This identity aligned closely with tales of technological revolutions, and the service propagated and supported these beliefs at all points during the evolution of contemporary unmanned systems. At times, force protection beliefs seemed like they might contradict this core commitment to technology, but the Air Force was adept at connecting the two, arguing that technology that distanced the human from the battlefield would not only create effectiveness but also decrease casualties in war. Preserving theories of strategic airpower has always been crucial for the Air Force, and the Air Force has consistently shown itself willing to adopt technologies to achieve this end.

Navy

The Navy in many ways mirrors the Air Force's love and embrace of technology. Ever since the transition to steam, the Navy's identity has been entangled with seaborne technology, and subsequently its love of capital ships, carriers, and even submarines has defined the Navy for a century.[69] However, there is a major difference between the Air Force and the Navy that has significant implications for the trajectory of unmanned systems. The Navy doesn't suffer from the service insecurity that creates the technology-grabbing incentives of the Air Force. Instead, the Navy has the opposite problem—an innate belief in itself and the utter superiority of the naval branch to master all domains. This leads to a Navy that adapts technology to the institution, instead of the institution adapting for the technology. As Daniel Lake concludes in his analysis of the Navy quest for technological superiority, "The strong traditional orientation of the Navy results in its incorporating new technologies into its existing institutional structure rather than transforming itself in response to technological change."[70]

The Navy's identity and innate self-confidence derive from its long history, one which the Navy tends to romanticize—particularly the golden age of seapower when dreadnoughts and capital ships dominated the seas and floating flotillas conducted diplomacy across the globe. Naval officers, no matter what their commissioning source, are all given an ideological foundation in the eighteenth-century theorist Mahan, from whom they learn that seapower plays a dominant role not only in warfare but also in trade, diplomacy, and international relations.[71] Whereas the Air Force espouses a theory of combat in which airpower can create strategic victory, the Navy's theory of victory goes one step further. Seapower not only creates decisive victory for ground campaigns but also ensures deterrence, builds alliances, and controls trade (and therefore ultimately a nation's economic prosperity).

Having such an expansive theory of the impact of seapower on military power obviously creates some practical difficulties for the Navy. Perhaps most importantly, there is an inherent tension (in budgets, manning, and seafaring resources) between the Navy's global presence mission and its lethality in conflicts.[72] As Montgomery McFate concludes in her analysis of the sociological foundations of the Navy's budget tendencies, these two missions lead to very different technologies for the Navy: "If the US Navy's core function 'is to kill people and break their toys,' then the Navy needs large surface combatant ships, underwater unmanned vehicles, hypersonic weapons and other highly lethal platforms and weapons. If, however, the Navy's core function is statecraft, then the Navy probably requires hospital ships, frigates, maritime civil affairs forces, and other platforms that enable coordination and collaboration with partner nations."[73] The Navy is therefore always struggling for resources to fulfill its Mahanian mandate of seapower across all domains of statecraft. Technology on the surface of the seas, underwater, and in the air compete within the Navy for supremacy, and the Navy's focus on large platforms significantly affects how it invests in emerging technologies.[74] The long lead time for development and construction of ships and submarines means that these premier platforms take up a large portion of research and development, procurement, and operation and maintenance budgets. Munitions and even aircraft have to be designed around the parameters of the ship and can be cut or surged in budgets based on changes in shipbuilding requirements or timelines.

For unmanned systems, this leads to a uniquely navalist trajectory, one in which the unmanned systems that survive support the existing hierarchy of branches (surface warfare, aviators, submariners) and (importantly) fit onboard platforms that have already been budgeted for, designed, and built decades prior. Unlike the Air Force, there was never significant debate about unmanned platforms replacing core manned platforms. Instead, the debate within the Navy was always about munitions and sensors—how investing in different unmanned enabling technologies would allow favored manned platforms to succeed within the inter-Navy budget fights. Sometimes this meant that there was more room for experimentation within the Navy as different branch identities could dip their toes into unmanned systems without threatening whole identities; but it also meant that it was harder to get some of these systems to maturity without advocates from within core warfare specialties.

The Navy dominated unmanned systems development and operations leading into World War II. However, the Navy didn't capitalize on its early lead in mines and torpedoes. Mine warfare was never a preferred tactic by the US Navy, and, barring short-term catalysts from conflict, mine development stagnated or atrophied throughout US naval history. Torpedoes, in contrast, are more accepted as part of traditional Navy missions. However, they struggled to compete with core missions and platforms. In the interwar years, torpedoes were viewed as competition for big-gun capital ships and aircraft. Absent an advocate within the Navy, torpedo factories were closed and the US Navy spent the first few years of World War II with a limited inventory of torpedoes with substantial design flaws. This decision to underresearch and underinvest in torpedoes had a knock-down effect on submarine warfare in World War II. US undersea warfare was far less capable than adversaries in the two world wars, and after World War II, the Cold War Navy was initially focused on aircraft carriers and airpower with limited interest in munitions that might enable undersea warfare. Torpedoes could have proliferated to aircraft or more firmly ensconced themselves within the core missions of submarines. However, Cold War torpedoes directly competed with new air-breathing cruise missiles and long-range, strategic ballistic missiles. Some of this edge was purely technological—torpedoes never developed the range or autonomy of cruise or ballistic missiles (though they also had much greater lethality against ships than missiles designed to hit from above the water instead of below). However, one of the major reasons torpedoes didn't proliferate to more missions à la their cruise or ballistic missile competitors was because torpedoes had no natural competitors in the Air Force or the Army. Absent service competition or support from a Navy warfighting community, torpedo research and development often stagnated. Meanwhile, early efforts to develop cruise missiles were initially galvanized by competition between the Air Force Matador and the Navy submarine-launched cruise missile, the Regulus.[75]

In the end, despite torpedoes' early lead, the Navy was more invested in missiles. For the Navy, the battle was never a choice between missiles and unmanned platforms or even between missiles and manned platforms but instead how three types of guided munitions—torpedoes, cruise missiles, and ballistic missiles—interacted with the powerful submarine, surface warfare, and aircrew communities within the Navy. Each of these munitions represented a different mission focus for the Cold War Navy, with implications

for both manned platforms and the Navy communities whose power fluctuated based on those platforms' dominance in Cold War Navy force structure.

The winner of the munition fight was the submarine community, and ballistic missiles in particular. This victory did not seem foreordained when the
competition between munitions began after World War II. Technologically,
cruise missiles appeared to be the right fit for an aircraft carrier–focused
US Navy. Unlike ballistic missiles, which had an unfortunate tendency to
damage or destroy aircraft carriers when launched from the deck, cruise
missiles could be safely launched from on board using much of the same
support system as aircraft. However, it was this similarity with airplanes
that significantly delayed cruise missile development within the Navy.
Cruise missiles—which looked very much like airplanes without pilots—
were a threat to aircraft carrier jets.[76] For example, Polmar and O'Connell
recount that when it came to the proliferation of the Regulus cruise missile, "naval aviators were not enthused" with missiles on the carrier because
"they already had manned aircraft to deliver nuclear bombs and operation
'unmanned aircraft' from carriers could cause flight deck handling and 'traf-
fic' problems."[77] It would take decades for cruise missiles to reach maturity
in the Navy, and ultimately, it was only when cruise missiles could be used
from multiple platforms and therefore support air, surface, and undersea
constituents that they became integrated within core Navy missions.

It is the ballistic missile that truly succeeds in the Navy internecine budget
fights. This is all the more remarkable because the ballistic missile initially failed resoundingly due to concerns that liquid-fueled ballistic missiles
would endanger the carrier. However, the Navy adapted ballistic missiles
to its own needs and developed a solid-propelled ballistic missile, launched
from a submerged submarine, which also created the aperture for a uniquely
navalist approach to nuclear deterrence in a Cold War US military. It was
helpful that the Navy decided to take a bite out of the nuclear deterrence
apple at the same time that Admiral Rickover was in an influence campaign
for new nuclear-powered submarines. As Graham Spinardi writes about the
period, "Although initially concerned that Polaris might divert resources
from his reactor work, Rickover soon came to see that their common purpose . . . could also have mutual benefit."[78] Importantly for Navy ballistic
missile development, Rickover needed the Polaris ballistic missile to justify
the expense of his new nuclear submarines.

The dominance of munitions in the naval unmanned trajectory makes the
Navy's very recent pivot toward unmanned platforms—including munitions

and to some extent surface ships—all the more remarkable. Throughout the decades in which ballistic and later cruise missiles began to proliferate through the Navy's arsenals, unmanned platforms made little progress. One particularly interesting case of an unmanned torpedo-firing antisubmarine helicopter, the Drone Anti-Submarine Helicopter (DASH),[79] revealed the challenge that individual leaders within the Navy faced when championing unmanned naval technologies that had no natural alignment with naval warfare identities. In this case, it was chief of naval operations Admiral Arleigh Burke who, frustrated with the accuracy of existing torpedoes (especially at long ranges), championed DASH to scout targets beyond the range of torpedoes, thus expanding the lethal range of the Navy's submarine-hunting capabilities.

Burke, a two-term chief of naval operations with significant power among the dominant surface warfare branch, pushed hard for DASH, condensing development timelines, securing and protecting the system's budget, and ordering their eventual deployment and use on Navy destroyers. But, despite strong top-down support, the program ran afoul of the Navy's branch cultures. Burke, a surface warfare officer, placed the drones on destroyers but never required the surface warfare community to learn how to operate them—nor did he assuage or co-opt aviators who hated the platform from its inception. As Tom Erhard explains in his analysis of the failed system, "DASH required a sailor with superior electronic and mechanical capabilities, but the most talented went to the carrier fleet, and the fatally short-changed training program could not recover."[80] Further, the aviator communities that might have had the skills to operate the system were frustrated with the notoriously difficult-to-operate drone. As Erhard details, "Carrier aviators contributed to the poisoned environment surrounding DASH operations, for the robot flew in 'their' airspace." His interviewees revealed that "[A]viators did not want to be in the air with that crazy thing.... all the pilots were afraid of flying into a drone."[81] Frustrated with the unmanned system, the aviation community proffered a manned alternative, and in 1966, when Burke came to the end of his two terms, the program was abandoned. The deputy chief of naval operations testified to the Senate that summer, recommending DASH be replaced with the manned alternative because "unfortunately, in robots you can't build judgment."[82]

Even Vietnam couldn't significantly invigorate support for unmanned platforms within the Navy. While the Air Force was investing in and experimenting with the Lightning Bug for battlefield reconnaissance and some suppression of enemy air defenses, the Navy aviation community had few unmanned resources. As Erhard explains, "Frustrated over wartime tactical reconnaissance shortfalls that they blamed on the intelligence community, the carrier navy asked for a jet-powered reconnaissance RPV in the mold of the Air Force's (NRO-funded) Lightning Bug series to complement the manned RA-5C Vigilante's strike reconnaissance mission."[83] The Navy, however, was not motivated by competition with the Air Force and largely ignored tactical use of unmanned aerial systems that did not directly support carrier survival.

Navy unmanned aircraft got a jolt with the appointment of a new secretary of the Navy, John Lehman, in 1981.[84] Lehman advocated for the Pioneer program after watching the use of UAVs in the Israeli-Arab wars and then experiencing the terrorist attack in Lebanon and subsequent shootdown of Navy jets by Syria. Lehman saw UAVs as part and parcel of a major expansion of the Navy to six hundred ships with a vision of assertive sea control.[85] Despite Lehman's efforts, when his term ended, so did the Pioneer program (though it saw more experimentation and adaptation in the Marines, which we will discuss below).[86]

Throughout the 1990s, the Navy unmanned trajectory stagnated. They invested in updates to some missile capabilities while dabbling in UAVs and underwater unmanned vehicles. However, these were lean budget years, and the Navy, with its expensive inventory of nuclear submarines and aircraft carriers, struggled to protect their aging fleets from significant reductions. With few combat threats to these fleets or its aviation component, the Navy didn't need to build unmanned systems for force protection, nor did it need to make significant unmanned innovations in domains under the water where it had lost its only real competitor, the Soviet Navy.

The 2000s heralded a new focus for the Navy that had survived the budget cuts of the 1990s. Conflicts in the Middle East and Navy support of counterterrorism operations in Afghanistan and Pakistan demonstrated once again the utility of naval-based aviation. Meanwhile, a rising China provided new impetus for naval growth—especially in the platforms (like aircraft carriers and submarines) that the Navy had always cared about. Technological narratives weaved into new doctrines like AirSea Battle and the third offset played into the Navy's core interests, and new calls for strike submarines and fifth-generation carrier-based aircraft echoed throughout DC think tanks and congressional corridors.

Despite the pivot toward Asia and the resurgence of naval technology in US military strategy, unmanned systems in the Navy lagged the Air Force and even the Army. The glut of resources that had gone into developing and operating UAVs like the Air Force's Predator and Reaper in Iraq and Afghanistan never diffused into the Navy. By the mid-2000s, there was a sense within the Navy that they were missing out on something about unmanned systems, though they struggled to articulate what exactly that was.[87] In an internal push to invest more in unmanned technologies, the Navy formed an unmanned office under their N9 directorate meant to overcome branch resistance to unmanned systems and push for these technologies across domains. There was a general perception that branch identities (especially aviators) were impeding Navy unmanned investment and leading to lopsided focus on domains (like subsurface) that were less likely to derail favored surface warfare programs. As former secretary of defense Leon Panetta recalled, "The Navy was more questioning about unmanned vehicles subsurface that could be used and developed. They thought it was worth looking at those capabilities—viewed as less of a threat to the Navy force structure."[88]

The 2000s once again saw civilian intervention in Navy unmanned systems investments. In particular, John McCain and his colleagues in the House and Senate Armed Services Committee pushed the Navy hard to develop a long-range unmanned carrier-based aerial vehicle program, Unmanned Carrier-Launched Airborne Surveillance and Strike (UCLASS). But UCLASS struggled, much as previous Navy unmanned efforts had, with no real narrative about what the system was supposed to do for the Navy and its core warfare constituencies. Unlike the Air Force, which often proffers to Congress a narrative about technology (and how that technology needs to supplement Air Force budgets), the Navy was content to let Congress spin its wheels, embarking on a long process of trials and experimentation that led to an unmanned refueler (and then intelligence resource) that neither the Navy nor Congress was particularly excited about.

The congressional hearings about UCLASS were revealing, if for no other reason than they highlighted the Navy and Congress's struggle to find a mission and purpose for carrier-based UAVs. Randy Forbes, who led the House discussion, opened the hearing by telling participants, "I believe the fundamental question we face is not about the utility of unmanned aviation to the future air wing, but the type of unmanned platform that the UCLASS program will deliver and specific capabilities this vital asset will provide the commander."[89] Congress didn't know what the Navy needed to use these UAVs for, but after seeing the Army and the Air Force adopting these systems en masse, the Navy's advocates in Congress believed the Navy needed to be prodded into figuring it out for themselves.

Also revealing was the wide array of missions for UCLASS presented to the House. Bryan McGrath, for example, trumpeted UCLASS as a revolutionary air strike asset, threatening even the future of other aircraft carriers. As he testified, "If the air wing of the future does not evolve in a way that enables the kind of unmanned strike that a truly capable UCLASS would bring, the aircraft carrier might indeed become obsolescent. If it becomes obsolete, the preponderant Navy that we field today that is the primary sustainer of the global system that is in place today will become far less powerful . . . [which] means a far less powerful and influential United States."[90] Meanwhile, Center for a New American Security expert Shawn Brimley presented a military revolutions argument for UCLASS, arguing that the revolutionary capabilities of long-range unmanned aerial strike could protect and preserve the now-endangered carriers from Chinese missile bases by preemptively striking these systems within the Chinese mainland. Other experts debated the purpose of a long-range aerial unmanned reconnaissance at all. There was no unifying narrative; whereas other services' unmanned trajectories were influenced by beliefs about technological theories of victory or casualty aversion, the Navy's strong sense of tradition and branch identities allowed the Navy to view unmanned systems as separate from other ideational fights.

The Navy's unmanned office in N9 had a short life; by the 2020s, it was disbanded. Ostensibly, the Navy's unmanned office had done what it was designed to do—overcome the infighting of warfare communities and galvanize support for unmanned systems. By the 2020s the Navy had a long-range aerial reconnaissance vehicle, a Navy-adapted Global Hawk.[91] Additionally, as of 2018 the Navy had its own underwater drone squadron[92] and an unmanned surface vehicle, the Sea Hunter.[93] Further, new plans for a larger Navy leaned heavily on unmanned surface vehicles (a huge dent in the 2021 research and development budget) to increase the size of the inventory and protect the vulnerable carrier from both budget and missile threats.[94] But the Navy never had a clear purpose for these unmanned systems, and even the squadrons of unmanned surface and underwater vehicles under development lack an operational vision. Meanwhile, the Navy embarked on an ambitious modernization of its missile inventory, including new long-range surface missiles and a modification to its nuclear ballistic missiles, but still lacks ship-launched missile systems (or launch platforms) that would be required for a significant conflict.

In the end, even after civilian pushes, budget cycles, and changes in the threat landscape, the Navy's unmanned trajectory remains largely the same. The Navy's core identity, its belief in itself, a large fleet, and the dominance of aircraft carriers, airplanes, and submarines, has not changed. Indeed, it hardly shifts even as services like the Air Force and the Army adapt their technology and unmanned investments to cater to dominant beliefs in military revolutions or public casualty aversion. For the Navy, the competition is always within itself, and the identity (and hierarchy) of surface warfare officers, aviators, and submariners plays a far greater role in Navy investment in unmanned systems than any policy entrepreneur or even combatant commander.[95] The Navy is relatively immune to shifts in beliefs; it largely ignored force protection beliefs, while using the military revolution narrative to invest in surface-based technology that it already preferred. Even as its leaders wrote books about network-centric warfare and pushed transformation, the service was relatively uninterested in unmanned systems—focusing instead on manned platforms with multi-decade-long lead times. If an unmanned system helped the Navy justify or improve one of these core platforms, then the Navy was likely to invest. However, when the systems challenged the platforms, the Navy tended to ignore or push back against the unmanned technology. The Navy, therefore, is an example of identity overtaking beliefs, of service and subservice allegiances dictating the adoption of technology over dominant ideas about warfare. This is perhaps only possible because the Navy was not as directly influenced by Vietnam or combat as the Army or even the Air Force and its ideas about technology are so tied to multi-decade-long acquisition and procurement that even successful ideas like revolutions in military affairs can only affect technological adoption on the margins.

Marines

The Marines, officially the ground arm of the Navy, have a distinct and entrenched identity. However, while the Navy is secure (almost smug) about their institutional existence, the Marines—as the smallest of the services—often worry about being overlooked. This is despite the Marines' hard-earned reputation as dogged fighters in expeditionary conflicts and a historical legacy in American military lore, going back hundreds of years. While the Marines were originally considered a police force of the Navy, meant to ensure shipboard security, their mission evolved over time to include raiding parties, amphibious assaults, and finally regular ground combatants. Their evolving missions sometimes leave them vulnerable to larger services, like the Army, that worry about the smaller service encroaching on their core mission and identity. This has led to a series of institutional coup attempts across the centuries. The Marines' ability to withstand these attacks has given the service a deep-seated distrust of the other services while also inculcating a strong sense of uniqueness and comradery. As former Marine Corps Lieutenant General Victor Krulak wrote, Marines "have learned through hard experience that fighting for the right to fight has often presented greater challenges than fighting their country's enemies."[96]

The Marines combat these challenges from other services by pursuing three strategies. First, the Marines teach every one of their members to be a Marine above any other identity, even their own—not a pilot, not a surface warfare officer, not even an infantryman. This identity-stripping process famously starts in boot camp when recruits are prohibited from using the pronoun "I" and receive the signature high and tight haircut—creating from the very first experience a sense of togetherness within the corps. The loyalty is to the unit, the corps, and the country, and these identity exercises stress the similarities between Marines as warriors, no matter what the mission, domain, or technology. As Sean Barrett describes the Marine ethos, "Whereas people in other services tend to identify more closely with their particular branch or those who do a similar type of work, Marines share a common identity as riflemen first and foremost. The Marine Corps lists as its first principle, which it claims helps define the cultural identity of Marines, "Every Marine a Rifleman. Every Marine—regardless of occupational specialty—is first and foremost a disciplined warrior." This identity has always led the Marine Corps to stress the human dimension in war and focus on the individual Marine—the rifleman, not weapons, technology, or systems—as its 'number-one priority.'"[97]

Second, Marines focus on differentiation from other services (especially the Army) by fostering an elite warrior identity. As Barrett explains, "In part due to this same institutional paranoia, the Marine Corps has cultivated an elite, almost mystical reputation and a culture of exceptionalism, demanding loyalty, sacrifice, discipline, frugality, and courage, which engenders a 'hunger for excellence' among its ranks. Marines have always insisted they are superior to other services."[98] While services like the Air Force may focus on exquisite technology and technological superiority to assert their uniqueness, the Marine Corps champions their superior individuals and a history of excellence on the battlefield.

Finally, as true identity entrepreneurs, the Marines have led a dedicated effort to promote their identity within Congress, in the public, and among other services. The Marines started this public affairs effort after World War II, when a restructuring of the DoD threatened the service. Barrett explains, "Feeling perpetually persecuted and under siege by the other services despite their effectiveness in the two world wars, the Marines became actively engaged in American society and politics as a means of survival, recruiting newspapermen, finding friends in Hollywood, sending veterans into politics, orchestrating congressional support, and building grassroots networks of support. In doing so, the Marines have leveraged their culture as a form of power and deployed it as a type of weapon or armor to protect themselves from external threats."[99] It is no coincidence that a slew of movies present the Marine warrior in battle or that Marine recruiting efforts play on this Hollywood mystique, for example, "looking for a few good men." The Marines were so effective at this public relations strategy that Truman is famously quoted as saying that the Marines "have a propaganda machine almost the equal of Stalin's."[100]

How, then, does the Marine identity affect the adoption of unmanned technology? Perhaps counterintuitively for a service that values the human and the warrior so highly, the Marines are one of the most adaptive to unmanned technology. This manifested in a series of different experimentation attempts with UAVs, dating back to the 1960s when the Marines first began to develop a small UAV able to take pictures from two thousand meters away.[101] The focus on small, organic-to-the-unit UAVs would be a constant in the Marine unmanned trajectory. Whereas other services had tendencies to gold-plate unmanned systems, create strategic missions for the platforms, or ignore them entirely, the Marines were interested in unmanned systems as a cheap augmentation of dull, dirty, and dangerous Marine missions. For the Marine, unmanned systems have always been about making the Marine rifleman better at their job, and therefore (unlike the Navy), there was never a need in search of a mission. As Erhard writes in his analysis of the role of Marine service identity in its investment in unmanned aerial systems, "In contrast to the Army, who first designed the SD series UAVs to perform a certain mission, then let requirements for manning and logistics evolve, the Marine Corps laid down its very slim manning and weight requirements and waited to see how much drone they could get."[102]

Despite Marine bottom-up experimentation with unmanned systems as early as the 1960s, their unmanned acquisitions programs never got strong support from the top-down Navy budgets. The obvious exception to this was the Lehman years and his vociferous support for UAVs to support both naval and marine missions. Under Lehman, the Marines were given the Pioneer UAV program and allowed large room for experimentation. They took this room and ran with it, forming a Mastiff platoon and then creating a concept of operations to employ the Pioneer from both land and amphibious ships.[103] The Marines went on to fly the most UAV sorties in the Gulf War (323 to the Navy's 83 and the Army's 46), and Marine commander General Walter Boomer "praised the Pioneer as that War's 'single most valuable intelligence collector.'"[104]

This is not to say that it was a seamless adoption of unmanned systems for the Marines. Marines may all be riflemen, but branch identities still affected the way in which unmanned systems were implemented. Aviators, in particular, did not at first support the implementation of the Pioneer UAV. As Erhard explains,

> A debate arose between Marine ground and air factions over resources and control of the asset. The crashes incensed Marine aviators because they were sending highly skilled aviation mechanics and pilots to the "toy airplane" project and they wanted control. If fielded, the "toy" would be flying in their airspace, and they did not want aviation illiterates flying the robot plane. The ground contingent had General Al Gray on their side, however, and he kept the focus on control by, and support to, the ground commander (later as terminology changed, he refused to call them UAVs, objecting to any name with "air" in it). Even in the Marine Corps, the UAV did not fit. The Marines proved remarkably flexible in adapting to the alien system, however. A genuine troop-level desire to make it work and high-level oversight overcame any divisions at this early stage. The new platoon commander, Bruce Brunn searched for a mix of aviation and ground troops that satisfied both sides, and the flexible Marine personnel system produced. "It was a great time to be a RPV commander," Brunn said, "because SECNAV [Lehman] was constantly pushing the program forward. RPVs would crash, and we would rebuild them as long as we could find a key piece of the original frame." Although

the Marines showed great flexibility in adapting to the UAV, the ground-air rift persisted and was not solved until over a decade later, revealing the "inbetween" nature of the UAV even in a culture as unified as the Marine Corps.[105]

In the end, the Marine identity defeated the branch identity. Marine aviators might have been reluctant at first to have artillerymen operating an unmanned system in their airspace, but in true Marine fashion, Marine aviators soon became part of the Pioneer units. The Marine can-do attitude overcame parochial, bureaucratic interests.

The Marines continued to adapt and innovate their own organic UAVs throughout the 2000s, standing up their own UAV units and staffing these with Marine pilots and unmanned operators. Further, as the DoD shifted its focus away from long campaigns of political attrition and back toward China and high-tech conflict, the Marines adapted once again. As RAND concluded, "With major investments in aviation, cyber, electronic warfare, and unmanned capabilities, the Marine Corps' culture may slowly be shifting toward a more technology-centric force. A China scenario highlights the utility of technology for the Marine Corps, as it would spur the Marine Corps to invest in its expeditionary advanced base concept, which utilizes F-35s. Unmanned systems, air defense, and sea control systems would be increasingly important, and new missions such as cyber and electronic warfare may even be worth sacrificing some attention to maneuver units."[106] Further, the Marine Force Design 2030 doubles down on unmanned systems, calling for a major acquisition package that includes both aerial and ship drones for logistics and resupply, intelligence, and small-unit strike support.

In the end, the Marine identity has large trickle-down effects to its relationship with technology, especially unmanned systems. First, the need to show the unique eliteness of the Marines (especially in comparison to the Army) allowed the Marines to be more flexible with their adoption of technology, especially without the luxury of the large ground contingent of forces resident in the Army. This may help explain why, despite the focus on ground combat and the strong identity of the Marine expeditionary warrior, the Marines have been early adopters and experimenters with unmanned systems. Further, because the Marines fall under the Navy, they retain an importance in future war with naval adversaries (like China) and therefore are more likely to see technological innovations like unmanned systems as part and parcel of their future brand of warfare—instead of a kind of warfare that they will be excluded from and therefore should resist. Finally, the Marines are not highly wedded to platforms for their service identity, which makes them more likely to buy and invest in weapons that fit their combat needs (not necessarily ones that fit their service identity). All these characteristics made it more likely that the Marines would adopt unmanned systems than some of the other services. However, because of their size, position under the Navy, and branch identities, Marine adoption of unmanned systems was never likely to be at the kind of scale or strategic focus as, for example, the Air Force. This makes the Marine a source of bottom-up innovation, a place for unmanned projects to receive the space for experimentation, even if the service is still vulnerable to the vagaries of Navy budgets.

Conclusion

When we first introduced the theory of this book, we spoke extensively about rational, capacity-based explanations and presented them as the status quo explanations for unmanned proliferation. But that was perhaps misleading. Our exploration of the extraordinary power of service identity to the ultimate unmanned trajectory within the US military shows that these identities were always going to be far more powerful and better explanations than any capacity-based reasoning. This is because the services are the ones that build capacity. Their identity comes a priori to almost anything else in a story about weapons development. Sometimes the services co-opt beliefs that help build their narrative, but usually they are resilient to changes in the beliefs that circulate through the Pentagon and apathetic to narratives pushed by policy wonks or civilian bosses.

For our story, service identities are often barriers to unmanned adoption; however, allegiance to these identities can create competition between the services and incentives to control the trajectory of unmanned systems. As John Deutch explained, "Service identities were barriers to adoption because they threatened the way they managed themselves; there are counterexamples, the issue of combined operations between Air Force and Army (airland operations)—once you saw that it was an operational necessity, then unmanned systems have an integral role in that concept (surveillance) and were more willing to invest in unmanned systems."[107] The two services most insecure were also the ones that were most likely to invest or adopt unmanned systems. For the Air Force, it was always about co-opting the technology to fit a narrative of strategic airpower and keeping it from a primarily ground support mission owned by the Army. For the Marines, it was about adapting technology to hone their effectiveness on the battle-field. And, without competing manned platforms, their biggest hurdle was not the Marine identity, but getting the resources from the Navy in the first place.

The Navy and the Army, as the most secure services, invested in unmanned systems that fit their beliefs about the dominance of their domains in warfare. For the Army, unmanned systems were never about replacing manned missions but instead about providing surveillance and force protection for the boots on the ground—thus ensuring public support that became so central to the Army's identity after Vietnam and the stand-up of the All-Volunteer Force. For the Navy, their journey with unmanned systems was about doing it their own way, not competing or emulating other services. Their hierarchical branches and the dominance of carrier concepts of seapower left them adrift for many years with unmanned platforms, even as competing branches allowed more space for innovation for munitions like ballistic and cruise missiles. The last few years have seen a surge in Navy support for unmanned systems, to include large platforms on the surface and underwater (a shift from the previous focus on only unmanned munitions). However, this surge seems to be related more to the Navy's overall focus on having a larger fleet and less on a clear theory of what these systems are supposed to do.

Finally, the strength of these identities over time demonstrates how diffi-cult it can be for policy entrepreneurs and nonorganic beliefs to influence weapons investments. Despite attempts by Congress and even the executive to temper the power of service identity in defense budgets, our story about unmanned systems illustrates the tenacity of these ideational allegiances to survive exogenous shocks, institutional realignments, and even domestic politics. In some ways, this makes the influence of beliefs like force pro-tection and military revolutions on contemporary unmanned investments all the more remarkable—their survival and influence despite entrenched identities is almost as puzzling a phenomenon as the rise of large drones in American arsenals. In our concluding chapter, we look at these processes in more detail, explaining how the messy role of beliefs and identity in hard security decisions has implications beyond unmanned systems.

Notes

1. Frank Kendall in discussion with the author, April 2020.
2. Thomas Erhard, "Unmanned Aerial Vehicles in the United States Armed Services: A Comparative Study of Weapons System Innovation" (PhD diss., Johns Hopkins University, June 2000).
3. The Space Force as a newly created service is still developing its identity and is largely not included in this study. However, the absence of this identity is discussed in our historical section when we introduce Eisenhower and his decision to place most space assets under a civilian organization, the National Reconnaissance Organization.
4. Builder, *The Masks of War*, 17.
5. Ibid., 19–20.
6. Ibid., 26–27.
7. Ibid., 34.
8. Ibid., 18.
9. Ibid., 21.
10. Ibid., 25.
11. Ibid., 29.
12. Ibid., 32.
13. Ronald O' Rourke, "Defense Primer: Department of the Navy," Federation of American Scientists, August 21, 2020, https://fas.org/sgp/crs/natsec/IF10484.pdf.
14. Mark Folse, "Marine Corps Identity from the Historical Perspective," War on the Rocks, May 2019, https://warontherocks.com/2019/05/marine-corps-identity-from-the-historical-perspective/; Leo Spaeder, "Sir, Who Am I? An Open Letter to the Commandant of the Marines," War on the Rocks, March 28, 2019, https://warontherocks.com/2019/03/sir-who-am-i-an-open-letter-to-the-incoming-commandant-of-the-marine-corps/.
15. Rebecca Jensen and Keil Gentry, *Identity Crisis between the Wars: How Doctrine Shaped the Marine Corps after World War I and Vietnam* (Quantico, VA: Marine Corps University, 2017)
16. S. Rebecca Zimmerman et al., *Movement and Maneuver: Culture and the Competition for Influence among the U.S. Military Services* (Santa Monica: RAND, 2019), https://www.rand.org/content/dam/rand/pubs/research_reports/RR2200/RR2270/RAND_RR2270.pdf.
17. We are not explicitly discussing the Space Force as it was only recently formed. However, the influence of space on Air Force identity is discussed as a key factor in Air Force unmanned investments.
18. Zimmerman et al., *Movement and Maneuver*, 19.
19. Ibid., 32.
20. Charles Nemfakos et al., *The Perfect Storm: The Goldwater-Nichols Act and Its Effect on Navy Acquisition* (Santa Monica, CA: RAND, 2010), https://www.rand.org/pubs/occasional_papers/OP308.html.
21. Zimmerman et al., *Movement and Maneuver*, 7.
22. Ibid., 12.
23. Ibid., 9.
24. Stine, *ICBM*.
25. Lonnquest and Winkler, *To Defend and Deter*, 226–233.
26. Erhard, "Unmanned Aerial Vehicles in the United States Armed Services."
27. Ibid., 274.
28. Ibid., 276.
29. Lock-Pullan, "How to Rethink War."
30. Peter Mansoor, "US Army Culture: 1973–2017," in *The Culture of Military Organizations*, d. Peter Mansoor and Williamson Murray (Cambridge: Cambridge University Press, 2019), 308.
31. Paul E. Menoher Jr., "Force XXI: Redesigning the Army through Warfighting Experiments," *Military Intelligence Professional Bulletin*, February 1996, https://fas.org/irp/agency/army/mipb/1996-2/menoher1.htm; Mark Hanna, "Task Force XXI: The Army's Digital Experiment," National Defense University Strategic Forum, July 1997, https://apps.dtic.mil/dtic/tr/fulltext/u2/a394389.pdf.
32. *Inside the Army* 9, no. 42 (October 20, 1997): 17.
33. *Inside the Army* 9, no. 13 (March 31, 1997): 1, 6–7.

34. Marc R. DeVore, "Reluctant Innovators? Inter-Organizational Conflict and the U.S.A.'s Route to Becoming a Drone Power," *Small Wars and Insurgencies* 31, no. 4 (2020): 711.
35. Wilson, "Senate Chairman Pushes Unmanned Warfare."
36. Erhard, "Unmanned Aerial Vehicles in the United States Armed Services."
37. Christopher Pernin et al., *Lessons from the Army's Future Combat Systems Program* (Santa Monica, CA: RAND, 2012), xxi, https://www.rand.org/pubs/monographs/MG1206.html.
38. General David Deptula in conversation with the author, April 20, 2020.
39. Former secretary of defense Leon Panetta in conversation with the author, April 2020.
40. Mansoor, "US Army Culture: 1973–2017," 299–318; Russell Frank Weigley, *The American Way of War: A History of United States Military Strategy and Policy* (Bloomington: Indiana University Press, 1977); Antulio Echevarria, *Toward an American Way of War* (Collingdale, PA: DIANE Publishing, 2004); Brian Linn, "The American Way of War Revisited," *Journal of Military History* 66, no. 2 (2002): 501; Bren Buley, *The New American Way of War: Military Culture and the Political Utility of Force* (Philadelphia: Routledge, 2007); Brian Linn, *The Echo of Battle* (Cambridge, MA: Harvard University Press, 2009).
41. Conrad Crane, *Avoiding Vietnam: The US Army's Response to Defeat in Southeast Asia* (Collingdale, PA: DIANE Publishing, 2002).
42. Zimmerman et al., *Movement and Maneuver*, 78.
43. Carl Builder, *The Icarus Syndrome: The Role of Air Power Theory in the Evolution and Fate of the U.S. Air Force* (New Brunswick, NJ: Transaction Publishers, 1996), 166.
44. Alex Abella, *Soldiers of Reason: The RAND Corporation and the Rise of the American Empire* (Boston: Houghton Mifflin Harcourt, 2009).
45. Zimmerman et al., *Movement and Maneuver*.
46. Builder, *The Icarus Syndrome*, 284.
47. Former secretary of defense Leon Panetta in conversation with the author, April 2020.
48. Builder, *The Icarus Syndrome*, 33.
49. As quoted in Timothy Schultz, *The Problem with Pilots* (Baltimore, MD: Johns Hopkins Press, 2018), 118.
50. J. C. Hopkins, *The Development of Strategic Air Command, 1946–1986 (The Fortieth Anniversary History)* (Offutt Airforce Base, Nebraska: Office of the Historian, Headquarters Strategic Air Command, 1986); William S. Borgiasz, *The Strategic Air Command: Evolution and Consolidation of Nuclear Forces, 1945–1955* (Westport, CT: Greenwood Publishing Group, 1996); Melvin Deaile, *Always at War: Organizational Culture in Strategic Air Command, 1946–62* (Annapolis, MD: Naval Institute Press, 2018); Melvin Deaile, "The SAC Mentality: The Origins of Organizational Culture in Strategic Air Command, 1946–1962" (PhD diss., University of North Carolina at Chapel Hill, 2007), https://doi.org/10.17615/xvz8-0261.
51. Schultz, *The Problem with Pilots*, 117.
52. Erhard, "Unmanned Aerial Vehicles in the United States Armed Services."
53. Mahnken, *Technology and the American Way of War since 1945*.
54. Craig Hannah, *Striving for Air Superiority: The Tactical Air Command in Vietnam*, no. 76 (College Station: Texas A&M University Press, 2002).
55. DeVore, "Reluctant Innovators?," 711.
56. Olsen, *John Warden and the Renaissance of American Air Power*.
57. "USAF Predator for the Future," Air Combat Command UAV CONOPs, as cited in Michael W. Kennedy, *A Moderate Course for USAF UAV Development* (Montgomery, Alabama: Maxwell Air Force Base: Air Command and Staff College, 1998).
58. Ibid.
59. Ibid.
60. Peter Grier, "New World Vistas," *Air Force Magazine*, March 1996, https://www.airforcemag.com/PDF/MagazineArchive/Documents/1996/March%201996/0396vistas.pdf.
61. *Global Engagement: A Vision for the Twenty-first Century Air Force*, 1996, https://apps.dtic.mil/dtic/tr/fulltext/u2/a318235.pdf.
62. Charles Dunlap, "Making Revolutionary Change: Airpower in COIN Today," *Parameters* 38, no. 2 (2008): 52–66, https://apps.dtic.mil/sti/pdfs/ADA490505.pdf.
63. As cited in Norman Schwartz and Suzie Schwartz, *Journey: Memoirs of an Air Force Chief of Staff* (New York, NY: Skyhorse Publishing, 2018), 296.
64. Ibid., 297.
65. Ibid., 297.

66. Ibid., 310.
67. General David Deptula in conversation with the author, April 20, 2020.
68. Former secretary of defense Leon Panetta in conversation with the author, April 2020.
69. Mark Regan Hagerott, "Commanding Men and Machines: Admiralship, Technology, and Ideology in the 20th Century US Navy" (PhD diss., University of Maryland, 2008); Roger W. Barnett, *Navy Strategic Culture: Why the Navy Thinks Differently* (Annapolis, MD: Naval Institute Press, 2009).
70. Daniel Lake, *The Pursuit of Technological Superiority and the Shrinking American Military* (New York: Palgrave, 2019), 149–150.
71. Ivan T. Luke, "Naval Operations in Peacetime: Not Just 'Warfare Lite,'" *Naval War College Review* 66, no. 2 (Spring 2013): 15
72. Montgomery McFate, "Being There: US Navy Organizational Culture and the Forward Presence Debate," *Defense and Security Analysis* 36, no. 1 (2019): 42–64.
73. McFate, "Being There," 52.
74. Lake, *The Pursuit of Technological Superiority and the Shrinking American Military*.
75. Polmar and O'Connell, *Strike from the Sea*; Werrell, *Evolution of the Cruise Missile*, 119.
76. Art and Ockenden, "The Domestic Politics of Cruise Missile Development, 1970–1980."
77. Polmar and O'Connell, *Strike from the Sea*, 44.
78. Spinardi, 79.
79. Friend, "Creating Requirements."
80. Erhard, "Unmanned Aerial Vehicles in the United States Armed Services," 320.
81. Ibid., 321.
82. Ibid., 322.
83. Ibid., 335.
84. Anthony R. Wells, *A Tale of Two Navies* (Annapolis, MD: Naval Institute Press, 2017).
85. Bill Keller, "The Navy's Brash Leader," *New York Times*, December 15, 1985, https://www.nytimes.com/1985/12/15/magazine/the-navy-s-brash-leader.html.
86. DeVore, "Reluctant Innovators?"
87. For example, a 2009 Chief of Naval Operations Strategic Studies Group detailed a potential trajectory for the Navy's unmanned investment across domains and called it "The Unmanned Imperative."
88. Former secretary of defense Leon Panetta in conversation with the author, April 2020.
89. Randy Forbes, "Comments during Hearing before the Subcommittee on Seapower and Projection of Forces," *Unmanned Carrier-Launched Airborne Surveillance and Strike (UCLASS) Requirements Assessment*, July 16, 2014, 2.
90. Bryan McGrath, "Comments during Hearing before the Subcommittee on Seapower and Projection of Forces," *Unmanned Carrier-Launched Airborne Surveillance and Strike (UCLASS) Requirements Assessment*, July 16, 2014, 11.
91. In a conversation with the author in April 2020, James Roche emphasized the Navy's need to develop its own systems, even inserting Navy-specific information technology requirements on its version of the Global Hawk that makes the Navy system incompatible with an Air Force logistical back end.
92. Joseph Trevithick, "The US Navy Has Created Its First Ever Underwater Drone Squadron," TheDrive.com, September 28, 2017, https://www.thedrive.com/the-war-zone/14733/the-us-navy-has-created-its-first-ever-underwater-drone-squadron.
93. Megan Eckstein, "Sea Hunter USV Will Operate with Carrier Strike Group, a SURFDEVRON Plans Hefty Testing Schedule," USNI News, January 21, 2020, https://news.usni.org/2020/01/21/sea-hunter-usv-will-operate-with-carrier-strike-group-as-surfdevron-plans-hefty-testing-schedule.
94. "Pentagon's Plan for a 500-Ship Navy Calls for More Unmanned Vessels," Maritime-Executive.com, October 7, 2020, https://maritime-executive.com/article/pentagon-s-plan-for-a-500-ship-navy-calls-for-more-unmanned-vessels.
95. B. P. Ponce, *Hollow Promises: The Problem of Culture and the Integration of New Technology into the Navy* (Fort Leavenworth, KS: Army Command and General Staff College, 2004).
96. Victor H. Krulak, *First to Fight: An Inside View of the U.S. Marine Corps* (Annapolis, MD: Naval Institute Press, 1984), 13–15.

97. Sean Barrett, *Always Faithful, Always Forward: Marine Corps Culture and the Development of Marine Corps Forces Special Operations Command* (Monterey, CA: Naval Postgraduate School, 2018), 55.
98. Ibid., 53.
99. Ibid.
100. Alan Rems, "A Propaganda Machine Like Stalin's," *Naval History Magazine* 33, no. 3 (2019), https://www.usni.org/magazines/naval-history-magazine/2019/june/propaganda-machine-stalins.
101. John Blom, Unmanned Historical Systems: A Historical Perspective, Occasional Paper 37 (Fort Leavenworth, KS: Combat Studies Institute Press, 2010).
102. Erhard, "Unmanned Aerial Vehicles in the United States Armed Services," 333.
103. Ibid., 361.
104. Bille Yenne, *Attack of the Drones: A History of Unmanned Aerial Combat* (Saint Paul, MN: Zenith, 2004), as cited in DeVore, "Reluctant Innovators?"
105. Erhard, "Unmanned Aerial Vehicles in the United States Armed Services," 351.
106. Zimmerman et al., *Movement and Maneuver*, xix.
107. Dr. John Deutch in conversation with the author, April 2020.

Chapter 6
Conclusion

I don't think it's an overstatement that this is a revolution of military affairs. The revolution is the conscious application of automated technology.

> —Colonel Eric Mathewson, Unmanned Aircraft Systems
> Task Force director

At first glance, the rise of unmanned systems seems like a simple story. These technologies make a military more capable, and therefore, states with more capacity will buy more and better unmanned systems. However, what this book shows is that nothing about the United States' current unmanned inventory is simple. In stark contrast to the narrative of inevitability that imbues many of the contemporary discussions of unmanned systems, ours is a story about contingency, where the roads not taken are as important as those traveled in explaining how we got to where we are today. It is a story that shows that beliefs about technologies often matter more than the technologies themselves; where the key protagonists are policy—or better yet, belief entrepreneurs—whose mission it is to see these beliefs reflected in budget lines; and how military services co-opt and resist these beliefs in accordance with their own entrenched identities. Finally, it is a story about how shifts arising from domestic politics and exogenous shocks create windows of opportunity for these beliefs to come to the fore, to compete, and ultimately to become embedded in doctrinal and budget decisions that create path dependencies for years to come.

This chapter serves a number of key purposes. It reviews where we have been so far—bringing the strands of our argument together in one clear statement. It then looks at contemporary puzzles in the world around us, focusing on the implications of investments in unmanned technologies for the Ukraine-Russia conflict in particular. Finally, it considers lessons learned from our story of the trajectory of unmanned systems for other technologies—including cyber, information technologies, space, and missiles.

The Hand Behind Unmanned. Jacquelyn Schneider and Julia Macdonald, Oxford University Press.
© Oxford University Press (2025). DOI: 10.1093/9780190064419.003.0007

Existing Arguments: Our Point of Departure

Before returning to our argument, it is important to emphasize that we do not directly disagree with many existing explanations for the rise of unmanned technologies. Dominant scholarly accounts point to the importance of state capacity and capability in developing and fielding unmanned systems. We agree that these are key determinants for countries that seek to develop complicated technology.[1] However, this literature doesn't ask *why* a state chooses to invest its scarce capabilities and capacity in unmanned technology over other possible technology investments (nor does it explain why there is variance within states in the types of unmanned systems that they develop). It tends to take as a given that unmanned technologies will enhance a state's military effectiveness and, because of this, that all states that are capable will want to develop and adopt unmanned systems. This would imply that the United States—clearly a highly developed state—would be investing heavily in unmanned technologies across the board. And yet we know from a quick glance at the current unmanned inventory and defense budget that investments have not been, and are not going to be, made equally across all services. Nor has the United States invested in some types of unmanned systems—mines, small unmanned aerial systems, loitering munitions—while it has overwhelmingly focused over the last two decades on over-the-horizon remotely controlled aerial platforms.

Existing accounts cannot explain these choices or the variation in investments between armed services. Why, for example, has the Air Force been the greatest advocate of unmanned aerial vehicles despite their threat to pilot identity? Why has the Army not invested more in unmanned autonomous ground vehicles, especially for high-risk missions in relatively uncluttered desert terrains? And why does the pattern of investment in the United States privilege platforms with precision strike that are controlled remotely over those that are more automated or less costly?

We also do not dwell on the arguments surrounding the legality of unmanned technologies under international law or the ethics of their use. There is a rich and interesting body of literature that explores the potential disconnect that remote warfare can create between the US public and the wars that it sanctions; the secrecy surrounding the use of unmanned aerial vehicles, especially the implications this holds for democratic accountability; and the status of drones in international law and the ethics of their use against other countries.[2] In our story, these debates and arguments are relevant insofar as they are drawn on instrumentally by belief champions to support their narratives for or against investments in unmanned technologies.[3] The details of these debates, including the legality of unmanned technologies, are not, however, the focus of this book.

Our Story

Masks of War: The Role of Service Identity in Technological Investments

As we saw in Chapter 5, perhaps the greatest determinant of when and why some unmanned technologies succeed while others fail is how the technology interacts with organizational incentives: namely, armed service identities and, at a more micro-level, occupational and warfare identities. It required significant external intervention—a war, Congress, or influential policy entrepreneurs—to overcome these identities. When the unmanned system threatened the service's identity, it was more likely to fail. When the unmanned system didn't have an advocate within the service, it was more likely to stagnate. Joint endeavors, often foisted by Congress in a top-down attempt to streamline or consolidate investment, were almost always unable to gain enough support from within services to survive over multiple budget years.

Sometimes occupational specialties that faced replacement by unmanned technologies resisted adoption, actively influencing the services through budgets, doctrine, and personnel choices.[4] This wasn't necessarily about manned versus unmanned platforms; munitions could also threaten favored platforms and lead to spirited campaigns against the technologies from within warfare communities. There are anecdotes throughout our story about pilots who balked at missiles and drones, surface warfare officers who disdained torpedoes or mines as tools of the weak, and infantry leaders who stalwartly defended boots on the ground over robotic options. However, sometimes unmanned systems benefited occupational identities, and then these communities became champions of systems that might have otherwise disappeared from defense budgets. The survival of weapons like submarine-launched ballistic missiles is a direct result of a competition between surface and subsurface warfare communities within the Navy.

So identity—whether occupational or service—does not uniformly stymie unmanned development. Competition between the services could also spur technological innovation—especially when Congress intervened to pit services against one another to develop munitions or platforms. This strategy was particularly useful in motivating the Air Force, whose consummate need to validate its existence meant that it was willing to adopt the technology – even if that technology countered its core service identity – as long as it meant that the Air Force could keep other services out. For example, the Air Force overcame its initial distaste for missiles to prevent the capabilities from falling into the hands of the Army.[5] Decades later, the Air Force's desire to control missions also led it to invest in and adopt remotely piloted aircraft at a far greater level than any of the other services—despite the power of the fighter pilot identity.

While service identities led to different unmanned technology trajectories within each service, in general, it created an incentive across services to focus on manned platforms, which services prized as core to their identity, over unmanned munitions and support equipment like bombs, missiles, communications, intelligence assets, or unmanned platforms. For example, despite early successes, the torpedo met significant resistance from the capital-ship Navy, which viewed the torpedo as a threat to its traditional structures. Decades later, the Navy continued its resistance to munitions that threatened operational identities, developing cruise missiles as a last resort to fend off ballistic missiles, which were incongruous with their platform of choice— the aircraft carrier. It wasn't until Admiral Hyman G. Rickover, who led the development of nuclear-powered submarines, saw the ballistic missile as a way to preserve and promote the submarine that ballistic missiles were embraced within the Navy budget (this also led to the innovation of liquid over solid propellants).

And while the Air Force's uneasy embrace of ballistic missiles has already been alluded to, the Air Force also put up a spirited resistance to cruise missiles in the 1970s, concerned that the munitions would decrease the chance that their new bomber, the B-1, would be funded. This resistance continued beyond missiles and into space, where the navigation satellites of the global positioning system were almost cut multiple times as the Air Force questioned the prioritization of space-based precision over pilot-directed laser-guided bombs. Finally, the Army's Future Integrated Combat System—which promised to link together lightly armored vehicles with drones and other support equipment—failed in part because the Army was more focused on the vehicles of the system and underfunded the network technology and unmanned sensors required to link the platforms together.

The Power of Narratives, Beliefs, and Policy Entrepreneurs

Given how entrenched these service identities are, the puzzle becomes when they *don't* explain weapons investments. In many cases, the reason that unmanned technologies survived organizational incentives and service identity bias was that they became part of a larger narrative. Narratives about the future of technology and beliefs about how technology might impact the future battlefield were key to convincing both service chiefs and Congress to preserve investments in technologies that might otherwise have been cut. In particular, we find that two core belief narratives dominated US defense discussions about technological investment over the last fifty years—those of military revolutions and of casualty aversion and force protection. For military revolution beliefs, technology is the primary agent of change. It is, therefore, the responsibility of the United States to harness the power of unmanned technologies to leapfrog adversaries by creating campaigns of speed, situational awareness, and decisive advantage. The second set of beliefs—derived primarily from the US experience in Vietnam and midgrade officers who dominated US defense thinking post–Cold War—looks to unmanned technologies to provide a technological solution to the constraints decision-makers believe are imposed by the American public's casualty intolerance.

Beliefs become more influential to policy when they are championed by enterprising individuals, policy entrepreneurs who become the hand behind unmanned systems—shaping, promoting, and combatting beliefs in order to build an arsenal that would not exist in a status quo ideational world. Policy entrepreneurs are most powerful when they come from within the services—for example, the Navy's Admiral Rickover pushing for ballistic missile submarines, the Air Force's General Curtis LeMay advocating for strategic bombers, General Bernard Schriever energizing the Air Force's ballistic missile development, or the Army's General Donn Starry creating the Corps 86 acquisition program to implement AirLand Battle.

The inherent power of within-service policy entrepreneurs benefited force protection beliefs, which developed organically from within warfighting communities, a deep-seeded idea about technology and the American way of war learned not from theory but instead through experience in Vietnam. These policy entrepreneurs were those officers like former chair of the Joint Chiefs of Staff Colin Powell—junior or midgrade officers during Vietnam—whose lessons learned about public support and casualty aversion influenced the doctrine and technologies they advanced as senior leaders. From the mid-1980s through to the early 1990s a series of military officers who had experience in Vietnam were promoted into positions of influence, including Marine General Anthony Zinni, Army General Carl E. Vuono, and Air Force General Ronald Fogleman. Often these individuals bucked entrenched service identities or sparred with competing subservice communities to spearhead unmanned technologies that they believed would insulate the American public from the human cost of war and advocated for unmanned systems focused on range, precision, and situational awareness with the aim to use technology to decrease the risk to American personnel.

In contrast, a second group of policy entrepreneurs, led by the Office of Net Assessment (ONA)'s Andy Marshall, leaned on civilian scholars, policy leaders, and networks of rising military officers to build and then propagate a theory of revolutions in military affairs. Marshall had little institutional power to influence beliefs. Instead of using normal tools of his influence, he had to use techniques and tools outside the standard defense decision-making process. To compensate for this lack of formal authority, he built a network of influencers who promoted a belief about technology and military revolutions throughout an entire defense community, infiltrating universities, think tanks, Congress, and military officers. With working groups, seminars, wargames, professional military education, and academic scholarship he built his own defense technocratic blob. He enlisted and mentored top thinkers, determined not only to influence the current secretaries of defense or top generals but also to build networks of ideational influence that could influence defense planning for a generation.

As a policy entrepreneur of a nonorganic belief, Marshall acted strategically when windows of opportunity opened to insert his ideas in the policy sphere, as well as to keep his ideas alive when those windows abruptly closed. In the 1990s, when the military revolutions narrative lost favor with the Clinton administration and budget cuts made new programs unlikely, he focused on maintaining investments in unmanned technologies of any kind, even if it wasn't the high-tech systems that he sought. In doing so, Marshall's disciples co-opted the force protection beliefs that dominated post–Cold War American military thought. The academic studies he commissioned became the intellectual foundations of the military revolutions narrative in years to come, allowing Marshall to influence beliefs even through periods of fraught relationships between the ONA and secretaries of defense.

Finally, occasionally there are individuals outside the services who can build powerful narratives and create networks of influence that circumvent service identity to influence technology investment choices from the outside in—Senator John McCain famously pushed the Navy and other services to evaluate their own biases, and President Dwight Eisenhower played an outsize role in shaping both the strategic arsenal of the United States after World War II and the United States' space capabilities. This internal group of military leader entrepreneurs, combined with the outside push of civilian policy entrepreneurs, led to sometimes conjoined narratives about technology to protect the force and create overwhelming technological victories. Together, these individuals created the crucial impetus by which beliefs about unmanned technology translated into policy and acquisition choices that could overcome status quo biases for service identity.

From the Outside In—the Exogenous Push and Pull on Technological Investments

So far, most of these lessons are about influencing defense technology processes from the inside. However, outside catalysts play a key role in shaping the trajectory of technology. Perhaps most obviously, wars drive innovation. They provide both an impetus and increased funding for technology, serving as immediate proving grounds for technology that might otherwise be favored (or rejected) by services.

This happened time and time again for unmanned systems. For example, torpedoes—which showed early promise—languished after World War I.[7] Faced with declining defense budgets, the Navy moved the systems to the bottom of their priority list, behind the new aircraft, submarines, and carriers the service was clamoring for. Only World War II forced the Navy to restart its torpedo development and introduce the far more capable Mark 24 and 28 torpedoes. Similarly, Vietnam drove tactical unmanned investments like unmanned aircraft and autonomous munitions, which were ignored under the peacetime dominance of the Strategic Air Command. More recently, 9/11 led to the armed remotely piloted aircraft, a phenomenon that has dominated unmanned system investments in the last two decades. In some cases, it wasn't just war that provided an exogenous shock to technology. While much of the Cold War was dominated by service identity and organizational competition, Sputnik and the nuclear threat created an impetus for investments in satellites and ballistic missiles that might have otherwise failed.[8]

Congress and, to some extent, the executive branch play important external catalyst (and confining/shaping) roles for technological development. While Congress is sometimes derided for retaining weapons that support industries within their representative districts, it often plays an important role in saving technology that would otherwise not be funded by the services. Congress and various presidential administrations, for example, can be credited for saving most of the ballistic and cruise missile technology that now exists in the US arsenal. It took civilian intervention, first from President Eisenhower's famous Solarium Commissions and then from senior civilians in Air Force research and development, to create a new protected entity within the Air Force focused solely on developing ballistic missiles.[9] Similarly, the Air Force's resistance to cruise missiles was largely the byproduct of the service's support for the B-1 bomber, and it was only because of Congress and the Carter administration that the Air Force's cruise missile program was saved.[10]

Congress and the executive branch also play an important role in influencing budget cycles, which serve as critical junctures for technological trajectories. Investments in research and development made in big-budget years create a path dependency for technologies during subsequent lean years. This can lead to suboptimal technological trajectories. Emerging technologies are more likely to be cut by the services during lean budget years, making outside intervention more important to saving technologies that are not otherwise preferred by the services. It also may leave the military with a glut of technology that isn't optimized for the current context. For example, the massive influx of defense spending during the Reagan years led to innovations in precision munitions but drawdowns in unmanned platforms, while post-9/11 conflicts inflated investment in remotely controlled unmanned platforms over long-range missiles or systems that may be useful in conflicts featuring more capable air defense systems.

Finally, the unmanned case suggests that the defense industrial base rarely innovates alone. There are very few cases of successful technological innovation started by the defense industrial base without a requirement from the Department of Defense (DoD). Instead, the tale of unmanned technologies is replete with examples of civilian innovations that fail to find customers within the military until a large exogenous shock forces the military to revisit (and sometimes resurrect) these technologies. This, understandably, decreases incentives for companies to produce technology that is not already explicitly requested by the DoD. The Predator is a rare exception; General Atomics invested in the system largely without a DoD push—a strategy that only succeeded because of the shock of 9/11.

The Ever-Present Intervening Variable: Technology Capacity and Capability

It would be impossible to conclude this story without recognizing the impact of technological developments on the trajectory of unmanned systems. At the same time that humans were making choices about unmanned technology, they were also making choices about communications, computing, propulsion, and stealth technology that affected the rise of unmanned technology. At some points in the story there were cultural incentives for more investment in unmanned technology and yet the technology wasn't quite ready to capitalize on that moment. Technological capability, therefore, does impact the diffusion and proliferation of these technologies. And certain technologies played an outsized role in explaining the contemporary US unmanned arsenal.

First and foremost, the development of radio communications enabled the initial use of remotely controlled technology. As those communications technologies matured, remote-controlled unmanned platforms became more and more feasible. It was the investment in unmanned systems in space, starting in the early 1960s, that enabled the remotely controlled unmanned aerial vehicles of today. Also, some unmanned platforms were more advantaged by these advances in communications technologies. While ground, sea surface, and air platforms became increasingly networked by communications within the electromagnetic spectrum, the undersea domain still struggled with underwater communications. This meant that remote-controlled systems could proliferate freely in the air, surface sea, and ground domains, but underwater systems had to be mostly autonomous.

Computing also shaped some of the choices about unmanned systems development. As alluded to within this brief discussion about communications technologies, the advent of digital information made it much easier for unmanned systems to be controlled but also to be able to collect, store, and transmit large quantities of information. The first impact was on guidance and the revolution of the microprocessor for developments in precision-guided missiles. These capabilities created by the microprocessor also enabled new unmanned platforms as the transition from vacuum tubes to microprocessors meant that platforms could be smaller or carry more sensors or munitions (a major development for unmanned aerial system experimentation in Vietnam). Computing went beyond missiles and microprocessors and enabled the network style of warfare undergirding military revolutions. As that technology became more mature, the autonomous, smart, and artificial intelligence–enabled systems envisioned by military revolutions beliefs also became more plausible. Even comparing some of the acquisitions programs of the early 2000s to today, these programs are less likely to fail because of a lack of datalink, storage, or data processing capabilities than they were even two decades ago (though bureaucratic divisions still fail many of these digitally focused programs).

Finally, advances in propulsion also affected the investment in cruise missiles and their alternative, unmanned strike platforms. The creation of the rocket in World War II followed by ballistic missiles during the Cold War was made possible by advancements in rocket propulsion. These advances made missiles a much more effective alternative to most of the unmanned platforms of the time. However, these advancements in propulsion largely flatlined as computing and communications drastically exploded, with exponential increases in capability for unmanned platforms from the digital revolution.

Technological advances certainly affected what kinds of unmanned systems were possible and cost effective. Separate investment decisions in microprocessors or computing had large trickle-down effects on the types of unmanned systems that could be used in conflict (and when ambitious unmanned visions had to be scrapped). These decisions are all part and parcel of the beliefs and identities that have shaped the trajectory of unmanned technology.

Implications: War in Ukraine

We began this book with a puzzle—why drones? Why a contemporary US unmanned arsenal dominated by remote-controlled expensive platforms like the Predator, Reaper, and Global Hawk? And, by contrast, why not more advancements in mines, missiles, and munitions? When we started this book, the contrast between the choices was a hypothetical: why did America envision one theory of victory over another? Our explanation has been a story about beliefs, identities, people, and exogenous shocks— how the American defense community built what they thought was the unmanned arsenal of the future. But now, as we come to the conclusion of our story, this is no longer a case of counterfactuals. Instead, battle-fields in Armenia, Ukraine, Syria, and Yemen are demonstrating a different unmanned reality. This reality is one not of large unmanned aerial plat-forms but instead of armies of small unmanned aerial systems, loitering munitions, kamikaze drones, and mines. How are these systems shaping the battlefield, and what effect will it have on the American unmanned arsenal?

Perhaps what has received the most public attention has been the domi-nance of small unmanned aerial systems—commercial, off-the-shelf drones; shoulder-launched systems; and longer-range kamikaze systems. These armies of drones are taking on a wide variety of missions: spotting for artillery fire, conducting intelligence and surveillance, providing logis-tics and communication relays, and dropping grenades and other smaller weapons.[11] Unlike American drones, which were designed to return home, controlled through a complex and expensive system of over-the-horizon communications technologies, these small systems are designed to create cheap mass. Who wins in the tit for tat of these systems versus electromag-netic jamming, counterdrones, artillery fire, machine guns, and netting is the side able to build the most systems at the cheapest cost and requiring the least labor.[12] The US arsenal of drones is poorly suited for this kind of warfare,[13] and Ukraine (as well as other nations like Iran and Turkey) have developed home-grown alternatives or adapted commercial options (like the Chinese DJI remote-controlled drones) to spur innovation to compensate for deficits within the US unmanned arsenals.

The difference between categories of munitions is increasingly difficult to parse as loitering munitions, kamikaze drones, glide bombs, and cruise missiles are used interchangeably on the modern battlefield. What all these systems have in common is that they are more adaptable and potentially more autonomous than previous munitions, allowing for dynamic and even independent retasking. Further, the evolution of these munitions has decreased the importance of platforms like aircraft and ships—a pragmatic reflection of a shortage of platforms on Ukraine's part and a vulnerability of ships and aircraft for the Russians. This comes in stark relief with the evolution of US unmanned systems in which munitions are often sacrificed for platforms or, in other cases, their use circumscribed by the need to conform with platforms preferred by the Navy or the Air Force. Set free of these cultural constraints on missile technology, Ukraine has, for example, adapted cruise missiles to launch from unmanned surface vehicles and truck-mounted systems.[14]

The Ukrainians' success highlights weaknesses in the US arsenal. Production lines for weapons like the Javelin and the Stinger were all but shut down before Russia invaded Ukraine.[15] The GLSDB—a 250-pound bomb delivered by rocket—originally received a hard pass from US military services.[16] To launch the antiship cruise missile Harpoon from land, the DoD had to draft a whole new emergency requirement.[17] Even the Army Tactical Missile System was cut at one point—its production only saved by a lifecycle contract extension. As Stacie Pettyjohn and Hannah Dennis conclude, the United States has been underinvesting in many munitions including "antiship and area-effects weapons" and is "not buying enough of these weapons" or "stockpiling enough precision-guided munitions (PGMs) for a protracted war."[18]

The conflict in Ukraine also highlights the importance of quantity of munitions over quality. Advances in air defenses—including networked radars, better processing capabilities, and smarter missiles—mean that more missiles and drones are destroyed before reaching their targets. Russian volleys on Ukraine, for example, in May of 2023, which included cruise missiles, hypersonic missiles, Shahed drones, and Orlan drones, cost potentially more than $1.4 billion with little to no impact on Ukraine's ability to sustain the war.[19] Whereas the early days of the precision-guided munition revolution suggested that wars could be won at long distances with cruise missiles and precision-guided bombs, today's wars demonstrate the difficulty that these weapons have in shifting the balance of power in a grueling war of attrition.

Drones and missiles may dominate public discussions about unmanned systems in Ukraine, but it is America's least favorite unmanned system, mines, that may have the largest effect on the war. Antitank mines, antipersonnel mines, and booby traps have been hallmarks of the Russian invasion, and Ukraine's counteroffensive has also used mines extensively.[20] The United States has even supplied Ukraine with mines, offloading decades-old antitank mines for the Ukrainian effort.[21] These mines, a relic of the US

Cold War arsenal, illustrate the effect of American apathy on the evolution of these unmanned technologies. Despite significant advances in drones and missiles, mines have not benefited from the digital revolution, and American inventories lack smarter automation or fuzing to make the systems either safer or more effective. Similarly, rudimentary naval mines have been used extensively in the Black Sea by both the Ukrainians and the Russians.[22] Like the terrestrial versions, technological advancements that benefited both drones and missiles have not diffused to naval mines. Meanwhile, minesweeping technology—for both land mines and naval mines—have received little attention from the United States, and attempts to invest in unmanned options have largely stagnated even as these systems' utility is demonstrated in Ukraine.

War is always an exogenous shock to technology, and the war in Ukraine will likely have as large an influence on the future US unmanned trajectory as the Arab-Israeli wars did in the 1970s. Most importantly, the conflict in Ukraine directly questions both of the beliefs about warfare that have shaped contemporary US investment in unmanned systems. According to the military revolution theory of war, unmanned technologies should enable rapid offensives and quick and decisive campaigns. However, what we have seen in Ukraine is how these systems have slowed down warfare. Drones are combatted with mesh netting over trenches and electromagnetic jamming. Offensives are stymied by low-tech land mines. Aircraft and naval vessels are destroyed by drones and swarms of cheap munitions. In short, the tit for tat of unmanned systems and counter-unmanned systems has incentivized bloody wars of attrition and not the short contests of technological overmatch foreshadowed by early believers in the information technology military revolution. As Franz-Stefan Gady writes of his observations of the Ukrainian battlefield:

> The advent of pervasive surveillance, these observers argue, has created a newly transparent battlefield. Ubiquitous drones and other technologies make it possible to track, in real time, any troop movements by either side, making it all but impossible to hide massing forces and concentrations of armored vehicles from the enemy. That same surveillance then makes sure that forces, once detected, are immediately hit by barrages of artillery rounds, missiles, and suicide drones. Sustaining any attempt at a breakthrough has become a most difficult proposition.[23]

These current trends in unmanned warfare also undermine some of the logics of technology associated with force protection beliefs. Unmanned systems didn't significantly remove the human from warfare or insulate forces from risk. Precise, more humane warfare didn't evolve from drones or missiles. Instead, as the costs of sustaining smart conflict increased, both Ukraine and Russia reverted to cheaper, less smart automated technologies like land and naval mines. Unmanned technologies created mechanisms for prolonged conflict without significantly increasing protection for the warfighter. War continued to be bloody, violent, and a battle of human will.

How influential will these lessons be for the United States' unmanned arsenal? Will these technologies proliferate to the United States, and how will they influence the power of beliefs and identities in shaping US unmanned inventories? There are a series of reasons why Ukraine will be a critical juncture for the US unmanned arsenal. First, the conflict reinvigorated munitions procurement, restarting production lines and injecting capital into parts of the defense industrial base that languished during the global war on terror. Further, Ukraine's unquenchable appetite for small drones catalyzed new entrants to the defense industrial base, building capacity within the global arms market that will have knockdown effects on American acquisition of the systems. Support for Ukraine has also saved a series of mothballed systems from the depths of defense budgets—antiship cruise missiles, air defense missiles, smart artillery, and ground-launched missiles are all thriving despite not being a priority for any of the armed services.

More importantly for our story about ideas and beliefs, the Ukraine war is shaping the next generation of military thinking, spurring rising officers and professional military education to explore new ways in which unmanned technology can shape modern warfare. Ideas about resilience, cost, and mass are emerging in defense strategies and service visions. The Space Force, for example, embarked on programs to create a civilian space reserve and to build lower-cost, higher-density low earth satellite programs. The Navy began its largest effort to invest in unmanned systems to date. This includes new missiles and munitions, but also unmanned surface vehicles and unmanned underwater vehicles. The Army is newly interested in ground-launched missiles and air defense missile technologies—two unmanned systems they largely ignored during counterinsurgency and abdicated to the Air Force during arms control missile reductions. Finally, the Marines fully embraced unmanned systems as part of their next evolution, calling for an inventory of unmanned systems for everything from communications to resupply to intelligence and strike.

Yet despite all of the reasons the conflict in Ukraine could significantly shift unmanned trajectories, there are also powerful reasons why the United States may not adopt the lessons learned from the war. First, service identity is remarkably sticky. Left to their own devices, the services struggle to prioritize acquisition programs beyond favored platforms. While today's conflicts demonstrate the primary role of unmanned technology as munitions, sensors, and stand-alone systems, the United States still has a status quo tendency to think about unmanned platforms as direct substitutions for manned systems. Even the use of "uncrewed" as the adjective de jure for unmanned systems implies a focus on unmanned as a direct substitution for manned systems instead of their utility as novel enablers to combined campaigns.

The United States has institutionalized these service identities in its arcane and labyrinthine acquisition and procurement process. Even acquisition innovations, built to create windows of opportunity for unmanned investment (like the Office of the Secretary of Defense and Defense Innovation Unit's replicator program), will struggle to scale without significant changes to bureaucratic incentives within the American defense apparatus. Ukraine's success has been a product of significant bottom-up innovation; this has fostered not only tactical experimentation with different weapons but also the rise of commercial firms able to quickly provide Ukraine with the technology its front-line troops want. This is still a struggle for the United States. There is very little room for bottom-up experimentation with new technology in the US military, and the DoD has struggled to put new defense technology companies on long-term procurement contracts.

Finally, one of the success stories for Ukraine's unmanned systems has been their investment in information technologies—software applications, flexible network architectures, and so forth—to link together unmanned systems to deliver flexible and mass effects. This remains a challenge for the United States, whose acquisition system is designed to acquire "things" and not constantly evolving information technologies. The difficulty the United States continues to have investing in joint information architectures, like the Joint All Domain Command and Control (JADC2), makes it extremely difficult to use unmanned systems in the ways that have been most effective in Ukraine.

Implications for Other Technologies

Though we do not develop a predictive theory in this book, our story about unmanned systems reveals broader lessons about when and why emerging technology succeeds or fails that can be tested and explored further by scholars.

First, the status quo for technologies is that they will succeed or fail based on how well they fit into a service's identity. To negate these biases, it is important to have the services compete against one another when necessary. Also, leaning on Congress and the executive branch can exert important top-down pressure when processes stagnate. Narratives, especially those propagated by successful policy entrepreneurs, can help overcome service biases. But perhaps the biggest implication of this research is that a general self-awareness within the American military of when technologies might inadvertently be set up to fail because they don't have organizations or individuals to spearhead them in the budget process could lead to better acquisition and development processes for emerging technology. There are also more specific implications for other emerging technologies.

Our exploration of unmanned systems suggests that cybertechnologies—especially those developed for joint missions—will need a champion to ensure their priority within defense budgets. Services are optimized to develop cybertechnologies that benefit their core identities, and yet the primary cyber force tasked with defending the nation is the Cyber National Missions Forces (CNMF), a joint organization subordinate to a joint functional command, Cyber Command.[24] Previous lessons from unmanned systems would suggest that carving out a joint organization to spearhead and defend the nation's mission (ransomware, defending elections, combating intellectual property theft, etc.) would be an uphill battle doomed to fail—fighting the services both for top talent to man billets and for control of cyber acquisition budgets. Indeed, for the last decade, the services have dominated this fight, allocating personnel to their missions first while almost all of the cyber acquisition budget passed straight through Cyber Command to the service cyber elements. Leading up to 2022, Cyber Command had budget oversight of only $600 million of an overall $40 billion DoD cyber/information technology budget (and even that small percentage was a significant increase from the $75 million Cyber Command was given authority over in 2018).[25]

But Cyber Command, led by the charismatic Army General Paul Naka-sone, waged a persistent and successful battle to influence legislation to pro-tect the CNMF and Cyber Command. Cyber Command was not only given authority (starting in 2024) for full budget control of the cyber/information technology portfolio but also elevated the CNMF to a subunified command, securing billets and budgets for the joint organization. How was Cyber Command able to do this? First, it crafted a narrative about its role, its mission's uniqueness compared to those of the services, and the need for greater authorities and budgets. In an effort colloquially known as persis-tent engagement, "Cyber Command published its strategic vision before the Trump National Defense Strategy or Defense Cyberspace Strategy."[26] Doing so preempted the cyber narratives in the DoD and National Secu-rity Council strategies and propagated the idea of a more forward-leaning and independent Cyber Command through academics, editorials, and pro-fessional military education.[27] This narrative benefited from the charisma of General Nakasone, who retained command of Cyber Command and the National Security Agency even while there was a revolving door of cyber leaders within the White House and in the position of secretary of defense. This campaign also benefited from a small pool of cyber leaders, many of whom served under Nakasone or Cyber Command before rotating to lead cyber service elements. By promoting within the Cyber Command cir-cle instead of from more traditional service hierarchies, Cyber Command retained influence within cyber service elements. Cyber Command also began a public-facing effort to advertise previously covert efforts, creating a new Twitter account to disseminate malware in real time, partnering with the (also charismatic) Chris Krebs at the Department of Homeland Secu-rity, and serving as a public face for a successful campaign to defend the US elections against Russian and foreign interference. The congression-ally mandated Cyberspace Solarium Commission also protected the Cyber National Mission Force, recommending and then spearheading the legisla-tion that ensured the force would remain funded and increasingly protected from the services. Of course, all these factors were helped by the services' relative disinterest in cyber versus more traditional missions, which allowed for Cyber Command to play a larger role in cyber budgets and control.

Related, and perhaps even more complicated, are investments in informa-tion technology and JADC2. The information technology structure of the military is already divided into armed service segments (the Air Force owns its network and data, etc.)—a phenomenon that makes basic information technology (and cybersecurity) upgrades difficult to implement across the DoD information technology network. It also leads the services to make very different decisions about applications, software development, and basic information technology practices. This creates complications for enterprise-wide initiatives through the Defense Information Systems Agency (DISA) and a natural jockeying between DISA and the armed service information technology organizations. It also means that JADC2—an effort to combine these networks to communicate and share data seamlessly across the services—is extremely difficult to implement. Joint organizations and initiatives are notoriously challenging, and the services' natural inclination to invest in platforms over infrastructure means that the ambitious information technology program faces multiple obstacles. However, lessons from the unmanned case about interservice competition suggest that Congress may incentivize the services to work together or prioritize information technology funding by threatening to allow one service to run the entire JADC2 program. Pitting, for example, the Air Force's air battle management system against the Navy's Project Overmatch may create an impetus for innovation where the status quo default is for stagnation.

The introduction of the Space Force complicates the role of service identity in investments in emerging technology. The power of service identities means there is a natural inclination for armed services, especially one that is new and concerned about its survival in the future, to push for technologies that create a novel role in space. While the Space Force is still developing its identity, the focus on offensive weaponry, warriors, and armed competition in space could lead to more investments in space-focused missions (e.g. space-based early warning, space weapons, etc.) over investments in intelligence, communications, and support for terrestrial missions. This was a large concern for President Eisenhower, who was apprehensive that service biases would harm space investments and needlessly cause a space race with Russia. To counteract these biases, Eisenhower placed almost all space capabilities under a civilian organization, the National Reconnaissance Organization. In the seventy years since, some would argue that Eisenhower's decision has underemphasized space capabilities in the defense budget and that the creation of the Space Force only corrects this imbalance. Further, by creating an armed service that must compete for resources, the new structure may lead to innovations in processes and personnel that lead to better acquisition and adoption of emerging technologies (as some would argue the Marines have exemplified). For space, the challenge will be surviving as an organization separate from the Air Force without leaning on identity biases that lead to less effective uses of emerging technology.

This research also has implications for the future of US conventional missile capabilities. Why hasn't the United States invested as much in hypersonic missiles as it probably should have? Why does the United States not have a larger, more sophisticated arsenal of conventional strike surface-to-surface (or even ship-to-ship or -shore) missiles? Part of the reason the nation has fallen behind countries such as North Korea and Iran in conventional strike options, and China in hypersonic missile options, is a product of context and the US-Soviet relationship in particular. Arms control agreements between the United States and the USSR limited much of the conventional strike arsenal; even cruise missiles (which were not explicitly a part of strategic arms control agreements) were used as part of a negotiating tactic for the United States in trying to limit nuclear arsenals.

Organizational interests also handicapped the US conventional strike development. The Army abdicated its stake in long-range missiles completely by the 1980s to the Air Force, which was happy to commandeer the mission from the Army but also was not interested in investing in missiles that didn't fit into the nuclear mission of Strategic Air Command, nor the conventional campaigns fought by Tactical Air Command. Meanwhile, the Navy (like the Air Force) lost interest in conventional strike missions that might threaten the bread-and-butter aircraft of aircraft carriers, while submarines, generally focused on strategic strike missions, were underprioritized as a conventional missile strike option. Even though missiles were a core part of technological narratives coming out of the DoD after the Cold War, these organizational and contextual complications meant that the focus was on missiles as munitions that could be carried by existing platforms like aircraft and destroyers. Even when leaders recognized the need for emerging technologies like hypersonic missiles, there were few service imperatives to invest in systems that didn't fall neatly into organizational niches and threatened the role of favored platforms.

Finally, this work reminds us that technology does not exist without human intervention. States cannot simply invest more in technology and expect it will lead to victory on the battlefield. Instead, how technology shapes the winners and losers in war is a result of the process by which organizations, individuals, and beliefs create and use that technology in the first place. For the US military, what this means for preparing for a future conflict with China or sustaining support to Ukraine against Russia is that the

United States cannot just increase defense budgets and expect that to translate into technological edge. Instead, the US military must focus as much on reforming the process of developing, acquiring, and implementing new technology as it does on fighting for larger budgets. A large part of this fight will be in re-examining the power of the armed services. It has been decades since the Goldwater-Nichols Act, the last major initiative to temper the role of service identity in defense budgets, and an evaluation of the success of these reforms is past due. Instead of building new services, the DoD should evaluate whether the current structure and power of the current division of services are best for military effectiveness. This will be a tall order that will require a Congress and executive branch willing to make difficult reforms at a time when civil-military relations are already strained. It will require new policy entrepreneurs from within and above the services, able to build compelling narratives for new weapons and concepts of operation.

Hopefully, the United States can do this without the push of war, because it is unclear that, if left to the status quo, the nation will be able to compete or win against an adversary on the level of China. In many of the emerging technologies that now dominate the discussion about future warfare—hypersonics, artificial intelligence, offensive space capabilities, even small drones—China looks to be an early leader. Meanwhile, after decades of conflict against terrorists and insurgents, the United States has an inventory of weapons ill-suited for the high-tech, primarily naval and air fighting in the Pacific.

A Theory of Unmanned Warfare for Future Trajectories

This book ultimately begs the question: what unmanned systems should the United States invest in? What does our story tell us about where the United States should focus its efforts? As we discussed in our first three chapters, how an unmanned system can contribute to warfare and in what ways it might make the biggest impact is largely a trade-off of different characteristics that investors believe are more or less likely to be important in future war. Persistence and precision, for example, often come at the expense of survivability, complicated logistical chains, and cost. On the other hand, unmanned munitions may come at a lower cost but require automation and may have limited persistence. Even characteristics that were once associated with revolutionary impacts of unmanned systems, for example, range or risk mitigation, have not shifted the balance of power in the Ukraine conflict. What the right unmanned trajectory is for the US military is less about which of these characteristics is most revolutionary and more about what kinds of unmanned characteristics are more or less important to the conflicts the United States believes it may be in over the next decade.

The beliefs and identities that we detailed in this book all proffer a different vision of the future of war. Beliefs in military revolutions are agnostic to the political stakes of a conflict and envision a theory of victory in which technological dominance creates massive overmatch and quick, offensive victories. Speed is the primary tenet of warfare for military revolutions and therefore they privilege unmanned systems that optimize speed and offensive strike. The metric of success is the time it takes to identify a target or threat and destroy it. For military revolution advocates, cost and risk are ancillary characteristics. Persistence and precision are important, but these characteristics are created not by singular systems loitering over targets but instead by building large networks of unmanned systems that together create an information advantage that is passed quickly to a strike platform or munition. Unmanned strike certainly decreases the time from sensor to shooter and can therefore fit the model. But slow, unsurvivable, remotely controlled systems like the Predator or Reaper are not natural fits for the future of warfare envisioned by those holding military revolution beliefs.

Those who align more closely with beliefs about casualty aversion and force protection are less worried about speed in warfare than they are the ability to decrease casualties on the battlefield. Maneuver and speed are great when they create battlefield conditions that decrease risk to friendlies (and we certainly saw these two beliefs elide during the 1990s), but in the end maneuver cannot win long wars of political attrition. For these wars of political attrition where ground troops are winning hearts and minds and establishing long-term presence, unmanned systems mitigate risk and may allow states to sustain conflict over time. They provide persistent coverage for ground troops and decrease the chance that pilots will be shot down. These systems don't need to be survivable, but they are more likely to be remotely controlled to decrease the chance of accidental fratricide or needless civilian loss of life.

It is not clear that either of these visions of future warfare align with the wars currently being fought. However, what this book suggests through its exploration of the last five decades of contemporary unmanned investment is that what really drives (or stymies) an unmanned trajectory is cost—whether that cost be economic or political. Economic cost was the demise of many unmanned systems, but when decision-makers prioritized mitigating political cost, unmanned systems flourished. The problem for those unmanned systems that prioritized political cost was that when the political stakes changed, these expensive unmanned systems were summarily cut.

This pattern of unmanned adoption suggested to us that the real theory of unmanned warfare in future conflict is not about range or maneuver or even precision but instead how both political and economic costs are privileged in two types of conflict. The first type of conflict, one in which the United States has found itself almost incessantly since the end of the Cold War, is a conflict of coercion in which aims are limited or in which conflict is a long-term competition of political will. The second kind of conflict is more akin to a near-peer high-stakes conflict in which the United States has vital interests. For that first kind of limited conflict, the United States prioritizes political cost. The second type of conflict is almost existential. In these cases the United States worries less about the political cost of loss of life and instead focuses on maintaining the economic cost of conflict. This suggests a potential theory of unmanned warfare that pivots away from range or even speed—perhaps the dominant characteristics of current discussions about unmanned advantage—and focuses instead on force protection and risk when states are competing short of war and on mass and economic cost when in near-peer war. Mitigating economic cost helps create mass and increase firepower, while mitigating political cost allows states to use weapon systems without disenfranchising domestic populations.

This leads us to a theory of unmanned warfare for two potential future trajectories. In low-stakes warfare or great power "coercion" in which the United States is concerned about escalation, unmanned systems can be built to mitigate political risk, investing in characteristics like risk mitigation, precision, and loiter or persistence above all other characteristics. This means investing in unmanned strike technologies that are potentially expensive and exquisite with costly sensors and remote operations by human controllers. It also means that economic cost in these conflicts is less important than political cost. Decision-makers should be willing to expend significant economic cost per platform to increase the persistence of these systems, their survivability, and the ability to control their actions. This is essentially the acquisition strategy of the global war on terror—in which focus on political cost substitutes machines for human risk.

By contrast, in existential, great power conflict these unmanned systems that mitigate political cost may have little utility. Because of the lack of quantity, these systems become high-demand, low-density assets that end up requiring protection by other assets. Therefore, unmanned systems in great power conflict must be designed to decrease economic cost, for example, serving as drones that soak up adversary missile inventories, providing redundant or resilient intelligence and communications assets, or massing for fire effects. To keep down logistical cost, most of these systems will need to be automated or autonomous (a characteristic that might limit the usefulness of strike platforms). They also must be expendable and therefore networks must be resilient enough to function and adapt when losing unmanned nodes. Because these systems are focused on economic cost, they provide advantages in missions that increase friction for the adversary, slowing down conflict, and creating survivability over time.

Both competition and conflict see an advantage for unmanned intelligence, surveillance, and reconnaissance missions—especially for sensors that can be deployed in mass and for low cost. As competition moves to conflict, these sensors need to become more and more replaceable with quick turnover times to replace destroyed sensors as well as resilient networks that are able to adapt to constant sensor replacements. Additionally, unmanned sensors that will succeed in conflict must be able to operate autonomously and have multiple modes of transmission to central repositories of information. Unmanned platforms that are high capability but also in low numbers (e.g. the Global Hawk) will be increasingly suboptimal as conflict intensifies. These platforms will have to be protected in a similar manner as manned alternatives and therefore will lose any revolutionary capability in conflict.

In the end, after taking this journey through the halls of the Pentagon, doctrinal revisions, professional military education syllabi, congressional hearings, think tank panels, and ultimately budget lines within the National Defense Authorization Act, we have become convinced that the rise of unmanned technology *should* be about cost. Instead of optimizing these systems for their ability to create cheap mass on the battlefield or to insulate decision-makers from political cost in wars we never meant to remain in, the American military has built unmanned systems that are expensive, direct substitutes for dying manned missions like dogfighting or carrier strike groups. All the while, we are still losing American forces on distant battlefields, and those very human losses don't seem to change public opinion about US foreign policy or budget spending.

Our aspiration in writing this book was to solve the puzzle about why the US military has invested in the kinds of unmanned systems that now make up its inventory. In the place of this puzzle, we hope we have given the US military the newfound power that comes with self-awareness to optimize its investments in unmanned systems—not free of ideas, beliefs, or identities but recognizing, correcting, and embracing these hugely influential forces to hone their unmanned inventory. The rise of unmanned systems is not inevitable. Instead, it is a reflection of human choices, organizational biases, and political shocks.

Notes

1. Gilli and Gilli, "The Diffusion of Drone Warfare"; Fuhrmann and Horowitz, "Droning On."
2. Michael Waltzer, "Targeted Killing and Drone Warfare," *Dissent*, January 11 2013; Audrey Kurth Cronin, "Why Drones Fail: When Tactics Drive Strategy," *Foreign Affairs* 92, no. 4 (2013): 44–54; Trevor McCrisken, "Obama's Drone War," *Survival: Global Politics and Strategy*, 55, no. 2 (2013): 97–122; Michael Aaronson and Adrian Johnson, eds., *Hitting the Target: How New Capabilities Are Shaping International Intervention* (London: Royal United Services Institute for Defence and Security Studies, 2013); Christian Enemark, *Armed Drones and the Ethics of War* (New York: Routledge, 2014); Stephanie Carvin, "Getting Drones Wrong," *International Journal of Human Rights* 19, no. 2 (2015): 127–141; Michael Boyle, ed., *The Legal and Ethical Implications of Drone Warfare* (New York: Routledge, 2017). For a more recent discussion of the broader implications of unmanned aerial systems see Michael Boyle, *The Drone Age: How Drone Technology Will Change War and Peace* (Oxford: Oxford University Press, 2020).
3. For further evidence supporting this claim of instrumentality see Stacie E. Goddard, "'The Most Humane of All Weapons': Discrimination, Air Power, and Precision Doctrine," *European Journal of International Security* 8, no. 4 (2023): 1–19. Goddard argues that rather than the norm of discrimination pushing the Air Force to adopt a precision doctrine, the Air Force co-opted the language of discrimination to justify their prior commitment to precision technology and doctrine.
4. Armin Granulo, Christoph Fuchs, and Stefano Puntoni, "Psychological Reactions to Human versus Robotic Job Replacement," *Nature Human Behaviour* 3, no. 10 (2019): 1062–1069.
5. Schultz, *The Problem with Pilots*.
6. Epstein, *Torpedo*, 10.
7. Ibid.
8. Robert Guerriero, *Space-Based Reconnaissance*.
9. Stumpf, *Minuteman*; Stine, *ICBM*.
10. Brown, *Flying Blind*; Art and Ockden, "The Domestic Politics of Cruise Missile Development 1970–1980," 359–413.
11. Emelia Probasco, "The Future of Drones in Ukraine," CSET, November 13, 2023, https://cset.georgetown.edu/article/the-future-of-drones-in-ukraine-a-report-from-the-diu-bravel-warsaw-conference/.
12. Ian Pannell and Allie Weintraub, "Inside Ukraine's Efforts to Bring an 'Army of Drones' to War against Russia," *ABC News*, September 15, 2023, https://abcnews.go.com/US/inside-ukraines-efforts-bring-army-drones-war-russia/story?id=103152130; David Axe, "Explosive Drones Are Everywhere in Ukraine. So the Infantry Head

Underground and Erect Screens," *Forbes*, November 16, 2023, https://www.forbes.com/sites/davidaxe/2023/11/16/explosive-drones-are-everywhere-in-ukraine-so-the-infantry-head-underground-and-erect-screens/?sh=599ee9df4b88.

13. There are reports that General Atomics offered to sell Predators to Ukraine for one penny (plus the cost of maintenance and logistics). None of these systems have been reported in theater as of yet so it is hard to be entirely clear whether Ukraine demurred or if these have been subject to export controls by the United States.

14. Howard Altman, "How the US Rushed Harpoon Anti-Ship Missiles to Ukraine," The Warzone, September 7, 2022, https://www.thedrive.com/the-war-zone/how-the-u-s-rushed-harpoon-anti-ship-missiles-to-ukraine.

15. Jacquelyn Schneider, "What It Will Take to Supply Ukraine for the Long Haul," *Wall Street Journal*, July 7, 2022, https://www.wsj.com/articles/what-it-will-take-to-supply-ukraine-for-the-long-haul-11657206373.

16. Nancy A. Youssef and Doug Cameron, "U.S. Expected to Send Ukraine Longer-Range Smart Bombs in Next Aid Package," *Wall Street Journal*, February 1, 2023, https://www.wsj.com/articles/u-s-expected-to-send-ukraine-longer-range-smart-bombs-in-next-aid-package-11675270504.

17. Heather Mongilio, "U.S. Sending Vehicle-Mounted Harpoon Launchers for Ukraine Coastal Defense," USNI News, June 15, 2022, https://news.usni.org/2022/06/15/u-s-sending-vehicle-mounted-harpoon-launchers-for-ukraine-coastal-defense.

18. Stacie Pettyjohn and Hannah Dennis, "Precision and Posture: Defense Spending Trends and the FY23 Budget Request," Center for a New American Security, November 17, 2022, https://www.cnas.org/publications/reports/precision-and-posture-defense-spending-tre.

19. "Night Attack on Ukraine Cost Russia at Least US $120 Million," *Pravda*, May 16, 2023, https://www.pravda.com.ua/eng/news/2023/05/16/7402408/. This means that states have to balance the need to create effects with these weapons with the cost to sustain enough volleys to make significant campaign progress.

20. Jen Kirby, "There Are Now More Land Mines in Ukraine than Almost Anywhere Else on the Planet," *Vox*, November 30, 2023, https://www.vox.com/world-politics/2023/11/30/23979758/ukraine-war-russia-land-mines-artillery-humanitarian-crisis.

21. John Ismay, "Pentagon Aid Packet to Ukraine Includes Decades-Old Anti-Tank Mines," *New York Times*, April 21, 2023, https://www.nytimes.com/2023/04/21/us/politics/ukraine-weapons-land-mines.html.

22. Lorenzo Tondo, "Sea Mines: The Deadly Danger Lurking in Ukraine's Waters," *The Guardian*, July 11, 2022, https://www.theguardian.com/world/2022/jul/11/sea-mines-ukraine-waters-russia-war-black-sea.

23. Franz-Stefan Gady, "How an Army of Drones Changed the Battlefield in Ukraine," *Foreign Policy*, December 6, 2023, https://foreignpolicy.com/2023/12/06/ukraine-russia-war-drones-stalemate-frontline-counteroffensive-strategy/.

24. Cyber Command Combined Action Group, "Beyond the Build: How the Component Commands Support the US Cyber Command Vision," *NDU Press*, January 1, 2016, https://ndupress.ndu.edu/Media/News/News-Article-View/Article/643106/beyond-the-build-how-the-component-commands-support-the-us-cyber-command-vision.

25. Mark Pomerlau, "Cyber Command Prepares to Gain Significant Budget Control," Fedscoop, March 4, 2022, https://www.fedscoop.com/cyber-command-budget-control-preparations-pom.

26. Jacquelyn G. Schneider, "Persistent Engagement: Foundation, Evolution, and Evaluation of a Strategy," Lawfare, May 10, 2019, https://www.lawfareblog.com/persistent-engagement-foundation-evolution-and-evaluation-strategy.

27. Michael P. Fischerkeller, Emily O. Goldman, and Richard J. Harknett, *Cyber Persistence Theory: Redefining National Security in Cyberspace* (Oxford: Oxford University Press, 2022); Paul M. Nakasone, "A Cyber Force for Persistent Operations," *Joint Force Quarterly* 92, no. 1 (2019): 10–14; Jacquelyn G. Schneider et al., *Ten Years In: Implementing Strategic Approaches to Cyberspace* (Newport, RI: US Naval College War Press, 2020).

Bibliography

Aaronson, Michael, and Adrian Johnson, eds. *Hitting the Target: How New Capabilities Are Shaping International Intervention*. London: Royal United Services Institute for Defence and Security Studies, 2013.

Abella, Alex. *Soldiers of Reason: The RAND Corporation and the Rise of the American Empire*. Boston: Houghton Mifflin Harcourt, 2009.

Acharya, Amitav. "How Ideas Spread: Whose Norms Matter? Norm Localization and Institutional Change in Asian Regionalism." *International Organization* 58, no. 2 (2004): 239–275.

Adams, Gordon, and Cindy Williams. *Buying National Security: How America Plans and Pays for Its Global Role and Safety at Home*. Philadelphia: Routledge, 2010.

Adamsky, Dima. *The Culture of Military Innovation: The Impact of Cultural Factors on the Revolution in Military Affairs in Russia, the US, and Israel*. Stanford, CA: Stanford University Press, 2010.

Adamsky, Dima P. "Through the Looking Glass: The Military-Technical Revolution and the Revolution in Military Affairs." *Journal of Strategic Studies* 31, no. 2 (2006): 257–294.

Adler, Emmanual. "The Emergence of Cooperation: National Epistemic Communities and the International Evolution of the Idea of Nuclear Arms Control." *International Organization* 46, no. 1 (Winter 1992): 101–146.

"Air Force, Navy to Develop Cruise Missile." *Aviation Week*, August 20, 1973.

Alberts, David S., John J. Garstka, and Frederick P. Stein. *Network Centric Warfare: Developing and Leveraging Information Superiority*. Washington, DC: Command Control Research Program, 2000.

Aldrich, John H., Christopher Gelpi, Peter Feaver, Jason Reifler, and Kristin Thompson Sharp. "Foreign Policy and the Electoral Connection." *Annual Review of Political Science* 9 (2006): 477–502.

Aldrich, John H., John L. Sullivan, and Eugene Borgida. "Foreign Affairs and Issue Voting: Do Presidential Candidates 'Waltz before a Blind Audience?'" *American Political Science Review* 83, no. 1 (1989): 123–141.

Allen, Thomas B., and Norman Polmar. *Rickover: Father of the Nuclear Navy*. Washington, DC: Potomac Books, 2007.

Altman, Howard. "How the US Rushed Harpoon Anti-Ship Missiles to Ukraine." The Warzone, September 7, 2022. https://www.thedrive.com/the-war-zone/how-the-u-s-rushed-harpoon-anti-ship-missiles-to-ukraine.

Alvis, Michael W. *Understanding the Role of Casualties in US Peace Operations*. Arlington, VA: Institute of Land Warfare, Association of the United States Army, 1999.

Arquilla, John, and David Ronfeldt. "Cyberwar Is Coming!" *Comparative Strategy* 12, no. 3 (1993): 141–165.

Art, Robert, and Stephen Ockenden. "The Domestic Politics of Cruise Missile Development." In *Cruise Missiles Technology, Strategy, Politics*, edited by Richard Betts, 359–413. Washington, DC: Brookings Institution, 1981.

Ash, Eric. "Casualty-Aversion Doctrine?" *Aerospace Power Journal* 14, no. 2 (Summer 2000): 2–3.

Atkeson, Brigadier General Edward B. "International Crises and the Evolution of Strategy and Forces." *Military Review* 55, no. 11 (1975): 47–55.

Atkinson, Rick. *Crusade: The Untold Story of the Gulf War*. New York: Houghton Mifflin Company, 1993.

Augier, Mie. "Thinking about War and Peace: Andrew Marshall and the Early Development of the Intellectual Foundations for Net Assessment." *Comparative Strategy* 32, no. 1 (2013): 1–17.

Axe, David. "Explosive Drones Are Everywhere in Ukraine. So the Infantry Head Underground and Erect Screens." *Forbes*, November 16, 2023. https://www.forbes.com/sites/davidaxe/2023/11/16/explosive-drones-are-everywhere-in-ukraine-so-the-infantry-head-underground-and-erect-screens/?sh=599ee9df4b88.

Baca, Glenn. *An Analysis of US Army Unmanned Ground Vehicle Strategy*. Cambridge/Carlisle, PA: MIT/US Army War College, 2012.

Bacevich, Andrew J. "Preserving the Well-Bred Horse." *National Interest* 37 (1994): 43–49.

Ball, Desmond. *Politics and Force Levels*. Berkeley: University of California Press, 1980.

Barnett, Roger W. *Navy Strategic Culture: Why the Navy Thinks Differently*. Annapolis, MD: Naval Institute Press, 2009.

Barrett, Sean. *Always Faithful, Always Forward: Marine Corps Culture and the Development of Marine Corps Forces Special Operations Command*. Monterey, CA: Naval Postgraduate School, 2018.

Bartels, Larry M. "Constituency Opinion and Congressional Policy Making: The Reagan Defense Buildup." *American Political Science Review* 85, no. 2 (1991): 457–474.

Baum, Matthew A. "Going Private: Public Opinion, Presidential Rhetoric, and the Domestic Politics of Audience Costs in US Foreign Policy Crises." *Journal of Conflict Resolution* 48, no. 5 (2004): 603–631.

Baum, Matthew A., and Philip B. K. Potter. "The Relationships between Mass Media, Public Opinion, and Foreign Policy: Towards a Theoretical Synthesis." *Annual Review of Political Science* 11 (2008): 39–65.

Beard, Edmund. *Developing the ICBM*. New York: Columbia University Press, 1976.

Berinsky, Adam J. *In Time of War: Understanding American Public Opinion from World War II to Iraq*. Chicago: University of Chicago Press, 2009.

Berkowitz, William R. "The Impact of Anti-Vietnam Demonstrations upon National Public Opinion and Military Indicators." *Social Science Research* 2, no. 1 (1973): 1–14.

Bertram, Volker. "Unmanned Surface Vehicles—A Survey." January 2008. http://citeseerx.ist.psu.edu/viewdoc/download?doi=10.1.1.462.1894&rep=rep1&type=pdf.

Black, Jeremy. "The Revolution in Military Affairs: The Historian's Perspective." *Journal of Military and Strategic Studies* 9, no. 2 (Winter 2006/2007): 98–102.

Blair, David. "Ten Thousand Feet and Ten Thousand Miles Reconciling Our Air Force Culture to Remotely Piloted Aircraft and the New Nature of Aerial Combat." *Air and Space Power Journal* 26, no. 4 (May–June 2012): 61–69.

Blaker, James R. *Understanding the Revolution in Military Affairs: A Guide to America's 21st Century Defense*. Washington, DC: Progressive Policy Institute, 1997.

Blom, John. *Unmanned Historical Systems: A Historical Perspective*. Occasional Paper 37. Fort Leavenworth, KS: Combat Studies Institute, 2010.

Blumenfeld, Aaron. "AirLand Battle Doctrine: Evolution or Revolution? A Look inside the U.S. Army." BA diss., Princeton University, 1989.

Bock, Joseph G. *The White House Staff and the National Security Assistant: Friendship and Friction at the Water's Edge*. Contributions in Political Science, no. 170. Westport, CT: Greenwood Press, 1987.

Boettcher III, William A., and Michael D. Cobb. "Echoes of Vietnam? Casualty Framing and Public Perceptions of Success and Failure in Iraq." *Journal of Conflict Resolution* 50, no. 6 (2006): 831–854.

Boot, Max. *War Made New: Technology, Warfare, and the Course of History, 1500 to Today*. New York: Penguin, 2006.

Borgiasz, William S. *The Strategic Air Command: Evolution and Consolidation of Nuclear Forces, 1945–1955.* Westport, CT: Greenwood Publishing Group, 1996.

Bousquet, Antoine. *The Scientific Way of Warfare: Order and Chaos on the Battlefields of Modernity.* New York: Columbia University Press, 2009.

Boyle, Michael. *The Drone Age: How Drone Technology Will Change War and Peace.* Oxford: Oxford University Press, 2020.

Boyle, Michael, ed. *The Legal and Ethical Implications of Drone Warfare.* New York: Routledge, 2017.

Bracken, Paul. "The Military after Next." *Washington Quarterly* 16, no. 4 (Autumn 1993): 157–174.

Branfill-Cook, Roger. *Torpedo: The Complete History of the World's Most Revolutionary Naval Weapon.* Annapolis, MD: Naval Institute Press, 2014.

Brose, Christian. *The Kill Chain: Defending America in the Future of High-Tech Warfare.* London: Hachette UK, 2020.

Brown, Michael. *Flying Blind: The Politics of the U.S. Strategic Bomber Program.* Ithaca, NY: Cornell Studies in Security Affairs, 1992.

Browne, Malcolm. "Invention That Shaped the Gulf War: The Laser-Guided Bomb." *New York Times,* February 26, 1991.

Buchholz, Arnold. "The Role of the Scientific-Technological Revolution in Marxism-Leninism." *Studies in Soviet Thought* 20, no. 2 (September 1979): 145–164.

Builder, Carl. *The Icarus Syndrome: The Role of Air Power Theory in the Evolution and Fate of the U.S. Air Force.* New Brunswick, NJ: Transaction Publishers, 1996.

Builder, Carl. *The Masks of War.* Baltimore, MD: Johns Hopkins University Press, 1989.

Buley, Bren. *The New American Way of War: Military Culture and the Political Utility of Force.* Philadelphia: Routledge, 2007.

Bush, George W. "A Period of Consequences." Speech, The Citadel, September 23, 1999. http://www3.citadel.edu/pao/addresses/pres_bush.html.

Calcara, Antonio, Andrea Gilli, Mauro Gilli, Raffaele Marchetti, and Ivan Zaccagnini. "Why Drones Have Not Revolutionized War: The Enduring Hider-Finder Competition in Air Warfare." *International Security* 46, no. 4 (2022): 130–171.

Calcara, Antonio, Andrea Gilli, Mauro Gilli, and Ivan Zaccagnini. "Will the Drone Always Get Through? Offensive Myths and Defensive Realities." 31, no. 5 *Security Studies* (2022): 791–825.

Campbell, John L. "Ideas, Politics, and Public Policy." *Annual Review of Sociology* 28 (2002): 21–38.

Campbell, Kenneth. "Once Burned, Twice Cautious: Explaining the Weinberger-Powell Doctrine." *Armed Forces and Society* 24, no. 3 (1998): 357–374.

Carter, Ralph G. "Senate Defense Budgeting, 1981–1988: The Impacts of Ideology, Party, and Constituency Benefit on the Decision to Support the President." *American Politics Quarterly* 17, no. 3 (1989): 332–347.

Carvin, Stephanie. "Getting Drones Wrong." *International Journal of Human Rights* 19, no. 2 (2015): 127–141.

Carvin, Stephanie, and Michael John Williams. *Law, Science, Liberalism and the American Way of Warfare.* Cambridge: Cambridge University Press, 2015.

Casey, Steven. *When Soldiers Fall: How Americans Have Confronted Combat Losses from World War I to Afghanistan.* Oxford: Oxford University Press, 2014.

Caverley, Jonathan D. *Democratic Militarism: Voting, Wealth, and War.* Cambridge: Cambridge University Press, 2014.

Caverley, Jonathan D. "The Myth of Military Myopia: Democracy, Small Wars, and Vietnam." *International Security* 34, no. 3 (2009): 119–157.

Chief of the Bureau [Ordnance] to President of the Torpedo Board. Report of the [Senate] *Select Committee on Ordnance and Warships.* Washington, DC: GPO, 1886.

Chittick, William O., Keith R. Billingsley, and Rick Travis. "A Three-Dimensional Model of American Foreign Policy Beliefs." *International Studies Quarterly* 39, no. 3 (1995): 313–331.

Chrysler Corporation Missile Division. *This Is Redstone.* Periscope Film LLC, 2012.

CIA. "USSR National Affairs Political and Social Developments." May 9, 1984. https://www.cia.gov/library/readingroom/docs/CIA-RDP90T00155R000500030019-6.pdf.

Clark, Wesley K. *Waging Modern War: Bosnia, Kosovo, and the Future of Conflict.* New York: Public Affairs, 2002.

Clinton, William J. *My Life.* New York: Alfred A. Knopf, 2004.

Cockburn, Andrew. *Kill Chain: The Rise of the High-Tech Assassins.* London: Picador Books: 2015.

Cohen, Eliot. "A Revolution in Warfare." *Foreign Affairs* 75 (March–April 1996): 37–54.

Cohen, Eliot A., Andrew J. Bacevich, and Michael J. Eisenstad. *"Knives, Tanks, and Missiles": Israel's Security Revolution.* Washington, DC: Institute for Near East Policy, 1998.

Cohen, Grant M. "Origins of U.S. Public Opinion for Drone Strikes: The Intersection of Elite Rhetoric, Media Coverage, and American Public Opinion, 2000–2015." PhD diss., University of Miami, 2018.

Coker, Christopher. *Humane Warfare.* New York: Routledge, 2001.

Coker, Christopher. *The Warrior Ethos: Military Culture and the War on Terror.* New York: Routledge, 2007.

Commission on Integrated Long-Term Strategy. *Discriminate Deterrence.* January 1988. https://apps.dtic.mil/dtic/tr/fulltext/u2/a277478.pdf.

Complete United States Infantry Guide for Officers and Noncommissioned Officers. Reprinted from government publications. Philadelphia: J. B. Lippincott, 1917.

"Conduct of the Persian Gulf Conflict: An Interim Report to Congress." April 24, 1992. https://apps.dtic.mil/sti/pdfs/ADA249445.pdf.

Converse, Elliot. *History of Acquisition in the Department of Defense: Rearming for the Cold War, 1945–1960.* Vol. 1. Washington, DC: Office of the Secretary, Historical Office 2012.

Cooper, Jeffrey R. "Another View of Information Warfare." In *The Information Revolution and National Security,* edited by Stuart J. Schwartzstein, 105–127. Washington, DC: Center for Strategic and International Studies, 1996.

Cooper, Jeffrey. *Another View of the Revolution in Military Affairs.* Carlisle, PA: US Army War College Strategic Studies Institute, 1994.

Correll, John T. "The Assault on EBO." *Air Force Magazine,* January 2013. https://www.airforcemag.com/PDF/MagazineArchive/Documents/2013/January%202013/0113EBO.pdf.

Correll, John T. "Casualties." *Air Force Magazine,* June 2003.

Correll, John T. "The Weinberger Doctrine." *Air Force Magazine,* March 1, 2014.

Crane, Conrad. *Avoiding Vietnam: The US Army's Response to Defeat in Southeast Asia.* Collingdale, PA: DIANE Publishing, 2002.

Cronin, Audrey Kurth. "Why Drones Fail: When Tactics Drive Strategy." *Foreign Affairs* 92, no. 4 (2013): 44–54.

Crowley, Jacelyn Elise. *The Politics of Child Support in America.* New York: Cambridge University Press, 2003.

Culpepper, Steven D. *Balloons of the Civil War.* Auckland, NZ: Pickle Partners Publishing, 2014.

Cyber Command Combined Action Group. "Beyond the Build: How the Component Commands Support the US Cyber Command Vision." NDU Press, January 1, 2016. https://ndupress.ndu.edu/Media/News/News-Article-View/Article/643106/beyond-the-build-how-the-component-commands-support-the-us-cyber-command-vision.

"DARPA and the Evolution of Unmanned." DefenseNews, April 2, 2017. https://www.defensenews.com/video/2017/04/02/darpa-and-the-evolution-of-unmanned-tech/.

Dauber, Cori. "Image as Argument: The Impact of Mogadishu on U.S. Military Intervention." *Armed Forces and Society* 27, no. 2 (2001): 205–229.

David, Charles-Philippe. "Policy Entrepreneurs and the Reorientation of National Security Policy under the G.W. Bush Administration (2001–04)." *Politics and Policy* 43, no. 1 (2015): 163–195.

de Mesquita, Bruce Bueno, Nils Petter Gleditsch, Patrick James, Gary King, Claire Metelits, James Lee Ray, Bruce Russett, Håvard Strand, and Brandon Valeriano. "Symposium on Replication in International Studies Research." *International Studies Perspectives* 4, no. 1 (2003): 72–107.

de Mesquita, Bruce Bueno, James D. Morrow, Randolph M. Siverson, and Alastair Smith. "An Institutional Explanation of the Democratic Peace." *American Political Science Review* 93, no. 4 (1999): 791–807.

Deaile, Melvin. *Always at War: Organizational Culture in Strategic Air Command, 1946–62.* Annapolis, MD: Naval Institute Press, 2018.

Deaile, Melvin. "The SAC Mentality: The Origins of Organizational Culture in Strategic Air Command, 1946–1962." PhD diss., University of North Carolina, 2007.

Department of Defense. *Annual Report to the President and the Congress by Donald Rumsfeld.* Washington, DC: US Government, 2002.

Department of Defense, Defense Airborne Reconnaissance Office. *Unmanned Aerial Vehicles Program Plan.* Washington, DC: US Government, 1994.

Department of the Army. *FM 100-5: Operations.* Washington, DC: US Government, 1976.

Deputy, Samuel. "Counterinsurgency and Robots: Will the Means Undermine the Ends?" Thesis, Naval War College, 2009.

Deterrence and Survival in the Nuclear Age (the Gaither Report of 1957), Joint Committee and Defense Production Congress of the United States, 94th Congress. Washington, DC: US Government Printing Office, 1976.

Devine, Troy E. *The Influence of America's Casualty Sensitivity on Military Strategy and Doctrine.* Maxwell AFB, AL: School of Advanced Airpower Studies Air University, 1997.

DeVore, Marc R. "Reluctant Innovators? Inter-organizational Conflict and the U.S.A.'s Route to Becoming a Drone Power." *Small Wars and Insurgencies* 31, no. 4 (2020): 701–729.

Dombrowski, Peter, and Andrew Ross. "The Revolution in Military Affairs, Transformation, and the Defense Industry." *Security Challenges* 4, no. 4 (Summer 2008): 13–38.

Dunlap, Charles. "Making Revolutionary Change: Airpower in COIN Today." *Parameters* 38, no. 2 (2008): 52–66.

Dyson, Stephen Benedict, and Thomas Preston. "Individual Characteristics of Political Leaders and the Use of Analogy in Foreign Policy Decision Making." *Political Psychology* 27, no. 2 (2006): 265–288.

Echevarria, Antulio. *Toward an American Way of War.* Collingdale, PA: DIANE Publishing, 2004.

Echevarria, Antulio J., and John M. Shaw. "The New Military Revolution: Post-Industrial Change." *Parameters* 22, no. 4 (Winter 1992–1993): 70–77.

Eckstein, Megan. "Sea Hunter USV Will Operate with Carrier Strike Group, as SURFDEVRON Plans Hefty Testing Schedule." USNI News, January 21, 2020. https://news.usni.org/2020/01/21/sea-hunter-usv-will-operate-with-carrier-strike-group-as-surfdevron-plans-hefty-testing-schedule.

Eichenberg, Richard C. "Victory Has Many Friends: US Public Opinion and the Use of Military Force." *International Security* 30, no. 1 (2005): 140–177.

Eichenberg, Richard C., Richard J. Stoll, and Matthew Lebo. "War President: The Approval Ratings of George W. Bush." *Journal of Conflict Resolution* 50, no. 6 (2006): 783–808.

Eikenberry, Karl W. "Take No Casualties." *Parameters* 26, no. 2 (1996): 109–118.

Enemark, Christian. *Armed Drones and the Ethics of War*. New York: Routledge, 2014.

Epstein, Katherine C. *Torpedo: Inventing the Military-Industrial Complex in the United States and Great Britain*. Cambridge, MA: Harvard University Press, 2014.

Erhard, Thomas. *Air Force UAVs: The Secret History*. Arlington, VA: Mitchell Institute Press, 2010.

Erhard, Thomas. "Unmanned Aerial Vehicles in the United States Armed Services: A Comparative Study of Weapons System Innovation." PhD diss., Johns Hopkins University, June 2000.

Everett, H. R. *Unmanned Systems of World War I and II*. Cambridge, MA: MIT Press, 2015.

Farrell, Theo G. "Culture and Military Power." *Review of International Studies* 24, no. 3 (Fall 1998): 407–416.

Farrell, Theo G. "Figuring Out Fighting Organizations: The New Organizational Analysis in Strategic Studies." *Journal of Strategic Studies* 19, no. 1 (Spring 1996): 122–135.

Farrell, Theo G., and T. Tariff. *The Sources of Military Change: Culture, Politics, Technology*. Boulder, CO: Lynne Rienner, 2002.

Fearon, James D. "What Is Identity (as We Now Use the Word)." Unpublished manuscript, Stanford University, 1999.

Feaver, Peter, and Christopher Gelpi. *Choosing Your Battles: American Civil-Military Relations and the Use of Force*. Princeton, NJ: Princeton University Press, 2005.

Finnemore, Martha. "Norms, Culture, and World Politics: Insights from Sociology's Institutionalism." *International Organization* 50, no. 2 (1996): 325–334.

Finnemore, Martha, and Duncan B. Hollis. "Constructing Norms for Global Cybersecurity." *American Journal of International Law* 110, no. 3 (2016): 425–479.

Finnemore, Martha, and Kathryn Sikkink. "International Norm Dynamics and Political Change." *International Organization* 52, no. 4 (1998): 887–917.

Fischerkeller, Michael P., Emily O. Goldman, and Richard J. Harknett. *Cyber Persistence Theory: Redefining National Security in Cyberspace*. Oxford: Oxford University Press, 2022.

Fitzgerald, Mary C. *Marshal Ogarkov on Modern War: 1977 to 1985*. Washington, DC: Center for Naval Analyses, 1987.

FitzSimonds, James R., and Jan M. Van Tol. "Revolutions in Military Affairs." *Joint Force Quarterly*, no. 4 (Spring 1994): 24–31.

Folse, Mark. "Marine Corps Identity from the Historical Perspective." War on the Rocks, May 2019. https://warontherocks.com/2019/05/marine-corps-identity-from-the-historical-perspective/.

Forbes, Randy. "Comments during Hearing before the Subcommittee on Seapower and Projection of Forces." *Unmanned Carrier-Launched Airborne Surveillance and Strike (UCLASS) Requirements Assessment*, July 16, 2014.

Ford, Matthew. *Weapons of Choice*. Cambridge: Cambridge University Press, 2017.

Franke, Volker. *Preparing for Peace: Military Identity, Value Orientations, and Professional Military Education*. New York: Greenwood Publishing Group, 1999.

Franks, Frederick. "Interview with Frederick Franks." *PBS Frontline: The Gulf War*, January 9, 1996. Full text available at http://www.pbs.org/wgbh/pages/frontline/gulf/oral/franks/5.html.

Freedman, Lawrence. *The Future of War: A History*. New York: Public Affairs, 2017.

Friend, Alice Hunt. "Creating Requirements: Emerging Military Capabilities, Civilian Preferences, and Civil-Military Relations." PhD diss., American University, 2020.

Frisbee, John L. "Project Aphrodite." *Air Force Magazine* 57 (1997): 57.

Fuhrmann, Matthew, and Michael C. Horowitz. "Droning On: Explaining the Proliferation of Unmanned Aerial Vehicles." *International Organization* 71, no. 2 (2017): 397–418.

Fulton, Robert. *Torpedo War and Submarine Explosions*. Chicago: Swallow Press, 1971 [reprint].

"The Future of Naval Capabilities." Center for Strategic and International Studies, last modified August 5, 2015. https://www.csis.org/events/future-naval-capabilities.

Future Security Environment Working Group. *The Future Security Environment*. October 1988. https://apps.dtic.mil/dtic/tr/fulltext/u2/a528366.pdf.

Gady, Franz-Stefan. "How an Army of Drones Changed the Battlefield in Ukraine." *Foreign Policy*, December 6, 2023. https://foreignpolicy.com/2023/12/06/ukraine-russia-war-drones-stalemate-frontline-counteroffensive-strategy/.

Gage, Douglas. "UGV History 101: A Brief History of Unmanned Ground Vehicle (UGV) Development Efforts." *Unmanned Systems Magazine* 13, no. 3 (1995): 1–10.

Gannon, Robert. *Hellions of the Deep: The Development of American Torpedoes in World War II*. University Park: Pennsylvania State Press, 1996.

Garfinkle, Adam. *Telltale Hearts: The Origins and Impact of the Vietnam Anti-War Movement*. New York: Macmillan, 1997.

Garrison, Jean. *Games Advisors Play: Foreign Policy in the Nixon and Carter Administrations*. College Station: Texas A&M University Press, 1999.

Gartner, Scott Sigmund. "The Multiple Effects of Casualties on Public Support for War: An Experimental Approach." *American Political Science Review* 102, no. 1 (2008): 95–106.

Gartner, Scott Sigmund, and Gary M. Segura. "War, Casualties, and Public Opinion." *Journal of Conflict Resolution* 42, no. 3 (1998): 278–300.

Gartner, Scott Sigmund, Gary M. Segura, and Bethany A. Barratt. "War Casualties, Policy Positions, and The Fate of Legislators." *Political Research Quarterly* 57, no. 3 (2004): 467–477.

Gartzke, Erik. "Kant We All Just Get Along? Opportunity, Willingness, and the Origins of the Democratic Peace." *American Journal of Political Science* 42, no. 1 (1998): 1–27.

Gartzke, Erik. "Preferences and the Democratic Peace." *International Studies Quarterly* 44, no. 2 (2000): 191–212.

Gates, Robert. "Speech on National Defense Strategy." September 29, 2008. https://insidedefense.com/document/gates-speech-national-defense-strategy.

Gelpi, Christopher, Peter Feaver, and Jason Reifler. "Iraq the Vote: Retrospective and Prospective Foreign Policy Judgements on Candidate Choice and Casualty Tolerance." *Political Behavior* 29, no. 2 (2007): 151–174.

Gelpi, Christopher, Peter Feaver, and Jason Reifler. *Paying the Human Costs of War: American Public Opinion and Casualties in Military Conflicts*. Princeton, NJ: Princeton University Press, 2009.

Gelpi, Christopher, Peter Feaver, and Jason Reifler. "Success Matters: Casualty Sensitivity and the War in Iraq." *International Security* 30, no. 3 (2006): 7–46.

Gentry, John A. "Casualty Management: Shaping Civil-Military Operational Environments." *Comparative Strategy* 30, no. 3 (2011): 242–253.

Cholz, Eugene, Harvey Sapolsky, and Caitlin Talmadge. *US Defense Politics: The Origins of Security Policy*. New York: Routledge, 2008.

Gibbons, William Conrad. *The US Government and the Vietnam War: Executive and Legislative Roles and Relationships, Part IV: July 1965–January 1968*. Princeton, NJ: Princeton University Press, 2014.

Gillespie, Paul. *Weapons of Choice: The Development of Precision Guided Munitions*. Tuscaloosa: University of Alabama Press, 2006.

Gilli, Andrea, and Mauro Gilli. "The Diffusion of Drone Warfare? Industrial, Organizational, and Infrastructural Constraints." *Security Studies* 25, no. 1 (2016): 50–84.

Giordano, Eric R. "The US Army and Non-Traditional Missions: Explaining Divergence in Doctrine and Practice in the Post-Cold War Era." PhD diss., Tufts University, 2003.

Gleannon, Michael. "Pre-empting Proliferation: International Law, Morality, and Nuclear Weapons." *European Journal of International Law* 24, no. 1 (2013): 109–127.

Global Engagement: A Vision for the Twenty-First Century Air Force. 1996. https://apps.dtic. mil/dtic/tr/fulltext/u2/a318235.pdf.

Gluth, Joseph V. "Is the Navy's Mine Warfare Posture Bankrupt?" Thesis, US Naval War College, 1991.

Goddard, Stacie E. "Brokering Change: Networks and Entrepreneurs in International Politics." *International Theory* 1, no. 2 (2009): 249–281.

Goddard, Stacie E. "'The Most Humane of All Weapons': Discrimination, Air Power, and Precision Doctrine." *European Journal of International Security* 8, no. 4 (2023): 1–19.

Goldgeier, James. *Not Whether but When: The US Decision to Enlarge NATO.* Washington, DC: Brookings Institute Press, 1999.

Goldstein, Cora Sol. "Drones, Honor, and War." *Military Review* 70 (2015): 70–76.

Goldstein, Judith, and Robert O. Keohane. *Ideas & Foreign Policy.* Ithaca, NY: Cornell University Press, 1993.

Goure, Daniel. "Is There a Military-Technical Revolution in America's Future?" *Washington Quarterly* 16, no. 4 (Autumn 1993): 175–192.

Graff, Henry F. *The Tuesday Cabinet: Deliberation and Decision on Peace and War under Lyndon B. Johnson.* Englewood Cliffs, NJ: Prentice-Hall, 1970.

Graham, Bradley. "Leader of US Troops in Mideast: An Unconventional Warrior." *Washington Post,* March 6, 1998.

Granulo, Armin, Christoph Fuchs, and Stefano Puntoni. "Psychological Reactions to Human versus Robotic Job Replacement." *Nature Human Behaviour* 3, no. 10 (2019): 1062–1069.

Gray, Colin S. *Strategy for Chaos: Revolutions in Military Affairs and the Evidence of History.* London: Frank Cass, 2002.

Gray, Colin. "Who's Afraid of the Cruise Missile?" *Orbis,* Fall 1977.

Grier, Peter. "New World Vistas." *Air Force Magazine,* March 1996. https://www.airforcemag. com/PDF/MagazineArchive/Documents/1996/March%201996/0396vistas.pdf.

Guerriero, Robert. *Space-Based Reconnaissance: From a Strategic Past to a Tactical Future.* Huntsville, AL: Army Space and Missile Defense Command, 2002.

Hacker, Barton C. "The Military and the Machines: An Analysis of the Controversy over Mechanization in the British Army, 1919–1939." PhD diss., University of Chicago, 1969.

Hafner-Burton, Emilie M. "The Behavioral Revolution and International Relations." *International Organization* 71 (Supplement 2017): S1–S31.

Hagerott, Mark Regan. "Commanding Men and Machines: Admiralship, Technology, and Ideology in the 20th Century US Navy." PhD diss., University of Maryland, 2008.

Hall, R. Cargill. "Reconnaissance Drones: Their First Use in the Cold War." *Air Power History* 61, no. 3 (2014): 20–27.

Halloran, Richard. "Officers Shaped by Vietnam Taking Control of the Army." *New York Times,* June 28, 1987. https://www.nytimes.com/1987/06/28/weekinreview/officers-shaped-by-vietnam-taking-control-of-the-army.html.

Halloran, Richard S. "US Buys Israeli Pilotless Planes." *New York Times,* last modified May 24, 1984. https://www.nytimes.com/1984/05/24/world/us-buys-israeli-pilotless-planes.html.

Halperin, Morton. "The Gaither Committee and the Policy Process." *World Politics* 13, no. 3 (1961): 360–384.

Hammond, William M. "The Press in Vietnam as Agent of Defeat: A Critical Examination." *Reviews in American History* 17, no. 2 (1989): 312–323.

Hanna, Mark. "Task Force XXI: The Army's Digital Experiment." National Defense University Strategic Forum, July 1997. https://apps.dtic.mil/dtic/tr/fulltext/u2/a394389.pdf.

Hannah, Craig. *Striving for Air Superiority: The Tactical Air Command in Vietnam.* College Station: Texas A&M University Press, 2002.

Harrison, Todd. *Defense Modernization Plans through the 2020s.* Washington, DC: CSIS, 2016.

Hartmann, Gregory. *Mine Warfare History and Technology.* White Oak Laboratory, PA: Naval Surface Weapons Center, 1975.

Hoffman, Erik P. "Review: Soviet Views of 'The Scientific-Technological Revolution.'" *World Politics* 30, no. 4 (1978): 615–644.

Hoffman, Frank. "A Second Look at the Powell Doctrine." War on the Rocks, February 2014. https://warontherocks.com/2014/02/a-second-look-at-the-powell-doctrine/.

Holsti, Ole R. *Public Opinion and American Foreign Policy.* Ann Arbor: University of Michigan Press, 2004.

Hopkins, J. C. *The Development of Strategic Air Command, 1946–1986 (the Fortieth Anniversary History).* Offutt, NE: Office of the Historian, Headquarters Strategic Air Command, 1986.

Hopkinsons, Francis. *The Battle of the Kegs.* Boston: Oakwood Press, 1866.

Horowitz, Michael, and Matthew Fuhrmann. "Droning On: Explaining the Proliferation of Unmanned Aerial Vehicles." *International Organization*, 71, no. 2 (2017): 397–418.

Houghton, David Patrick. "Historical Analogies and the Cognitive Dimension of Domestic Policymaking." *Political Psychology* 19, no. 2 (1998): 279–303.

House Committee on Appropriations, 65th Congress, Third Session. *Regarding Torpedoes for Coastal Defense.* October 3–9, 1918.

Hovland, Carl, Irving L. Janis, and Harold H. Kelley. *Communication and Persuasion.* Westport, CT: Greenwood Press Publishers, 1953.

Howze, Hamilton. "Vietnam: An Epilogue." *Army*, July 1975: 12–16.

Huisken, Ronald. *The Origin of the Strategic Cruise Missile.* New York: Praeger, 1981.

Hurley, Alfred F. *Billy Mitchell: Crusader for Air Power.* Bloomington: Indiana University Press, 1964.

Hyde, Charles K. *Casualty Aversion: Implications for Policy Makers and Senior Military Officers.* Newport, RI: Naval War College, 2000.

Ismay, John. "Pentagon Aid Packet to Ukraine Includes Decades-Old Anti-Tank Mines." *New York Times*, April 21, 2023. https://www.nytimes.com/2023/04/21/us/politics/ukraine-weapons-land-mines.html.

ablonsky, David. "U.S. Military Doctrine and the Revolution in Military Affairs." *Parameters* 24, no. 3 (Autumn 1994): 18–36.

Jackman, Anna H. "Rhetorics of Possibility and Inevitability in Commercial Drone Tradescapes." *Geographica Helvetica* 71, no. 1 (2016): 1–6.

Jensen, Rebecca, and Keil Gentry. *Identity Crisis between the Wars: How Doctrine Shaped the Marine Corps after World War I and Vietnam.* Quantico, VA: Marine Corps University, 2017.

Jentleson, Bruce W. "The Pretty Prudent Public: Post Post-Vietnam American Opinion on the Use of Military Force." *International Studies Quarterly* 36, no. 1 (1992): 49–73.

Jertleson, Bruce W., and Rebecca L. Britton. "Still Pretty Prudent: Post-Cold War American Opinion on the Use of Military Force." *Journal of Conflict Resolution* 42, no. 4 (1998): 395–417.

Jervis, Robert. *Perception and Misperception.* Princeton, NJ: Princeton University Press, 1976.

Jo, Dong Joon, and Erik Gartzke. "Determinants of Nuclear Weapons Proliferation." *Journal of Conflict Resolution* 51, no. 1 (2007): 167–194.

Johansen, Rino Bandlitz, Jon Christian Laberg, and Monica Martinussen. "Military Identity as Predictor of Perceived Military Competence and Skills." *Armed Forces and Society* 40, no. 3 (2014): 521–543.

Johnson, Daniel Isaac. *Death and Doctrine: U.S. Army Officers' Perceptions of American Casualty Aversion 1970–1999.* College Station, TX: Texas A&M University, 2010.

Johnson, Kent. "Political-Military Engagement Policy: Casualty Avoidance and the American Public." *Aerospace Power Journal* 15, no. 1 (Spring 2001): 98–105.

Johnson, Lyndon B. "Remarks in Chicago before the National Association of Broadcasters." In *The Public Papers of the President of the United States, 1968*, edited by Lyndon B. Johnson. Washington, DC: US Government Printing Office, 1969.

Joint Chiefs of Staff. *National Military Strategy of the United States.* 1992. https://history. defense.gov/Portals/70/Documents/nms/nms1992.pdf?ver=2014-06-25-123420-723.

Joyce, Adam. "The Micropolitics of 'the Army You Have': Explaining the Development of U.S. Military Doctrine after Vietnam." *Studies in American Political Development* 26, no. 2 (2012): 180–204.

Jungdahl, Adam M., and Julia M. Macdonald. "Innovation Inhibitors in War: Overcoming Obstacles in the Pursuit of Military Effectiveness." *Journal of Strategic Studies* 38, no. 4 (2015): 467–499.

Jurkovich, Michelle. "What Isn't a Norm? Redefining the Conceptual Boundaries of 'Norms' in the Human Rights Literature." *International Studies Review* 22, no. 3 (2020): 693–711.

Kaag, John, and Sarah Kreps. *Drone Warfare.* Hoboken, NJ: John Wiley & Sons, 2014.

Kaempf, Sebastian. "US Warfare in Somalia and the Trade-Off between Casualty-Aversion and Civilian Protection." *Small Wars and Insurgencies* 23, no. 3 (2012): 388–413.

Kahl, Colin. "Constructing a Separate Peace: Constructivism, Collective Liberal Identity, and Democratic Peace." *Security Studies* 8, no. 2–3 (1998): 94–144.

Kahl, Colin. "In the Crossfire or the Crosshairs? Norms, Civilian Casualties, and US Conduct in Iraq." *International Security* 32, no. 1 (2007): 7–46.

Karman, Theodore von. *Prophecy Fulfilled: "Toward New Horizons" and Its Legacy Fulfilled.* Maxwell, AL: Air Force History and Museums Program, 1994.

Katzenstein, Peter J. *The Culture of National Security: Norms and Identity in World Politics.* New York: Columbia University Press, 1996.

Katzenstein, Peter J., and Nobuo Okawara. "Japan's National Security: Structures, Norms, and Policies." *International Security* 17, no. 4 (1993): 84–11.

Katznelson, Ira. "Periodization and Preferences." In *Comparative Historical Analysis in the Social Sciences,* edited by James Mahoney and Dietrich Rueschemeyer, 270–302. Cambridge: Cambridge University Press, 2004.

Keane, John F., and Stephen S. Carr. "A Brief History of Early Unmanned Aircraft." *Johns Hopkins APL Technical Digest* 32, no. 3 (2013): 558–571.

Keaney, Thomas A., and Eliot A. Cohen. *Gulf War Air Power Survey: Weapons, Tactics, and Training and Space Operations.* Vol. 4. Arlington, VA: Office of the Secretary of the Air Force, 1993.

Keller, Bill. "The Navy's Brash Leader." *New York Times,* December 15, 1985. https://www. nytimes.com/1985/12/15/magazine/the-navy-s-brash-leader.html.

Kelley, M. R. "The Terrier: A Capsule History of Missile Development." *APL Technical Digest* 4, no. 6 (1965): 18–26.

Kelly, John H. "Lebanon: 1982–1984." In *U.S. and Russian Policymaking with Respect to the Use of Force,* edited by Jeremy R. Azrael and Emil A. Payin, 85–104. Santa Monica, CA: RAND Corporation, 1996.

Kendall, Frank. "Exploiting the Military Technical Revolution: A Concept for Joint Warfare." *Strategic Review* 20, no.2 (Spring 1992): 25.

Kennedy, Michael W. *A Moderate Course for USAF UAV Development.* Maxwell AFB, AL: Air Command and Staff College, 1998.

Kennett, Lee. *The First Air War: 1914–1918.* New York: Simon and Schuster, 1999.

Kier, Elizabeth. "Culture and Military Doctrine." *International Security* 19, no. 4 (Spring 1995): 65–93.

Kimball, Jeffrey P. "The Stab-in-the-Back Legend and the Vietnam War." *Armed Forces and Society* 14, no. 3 (1988): 433–458.

King, William. *Torpedoes: Their Invention and Use from the First Application to the Art of War to the Present Time.* Washington, DC: n.p., 1866.

Kingdon, John. *Agendas, Alternatives, and Public Policies.* Boston: Little, Brown, 1984.

Kinnard, Douglas. *War Managers.* Annapolis, MD: Naval Institute Press, 1977.

Kirby, Jen. "There Are Now More Land Mines in Ukraine than Almost Anywhere Else on the Planet." *Vox*, November 30, 2023. .https://www.vox.com/world-politics/2023/11/30/23979758/ukraine-war-russia-land-mines-artillery-humantarian-crisis.

Kissinger, Henry, to Richard Nixon. "Impact of American Casualties on Your Vietnam Policy." June 21, 1969, ND8–1. Casualties, Box 26, Nixon Papers.

Klarevas, Louis J. "Trends: The United States Peace Operation in Somalia." *Public Opinion Quarterly* 64, no. 4 (2000): 523–540.

Klotz, Audie. *Norms in International Relations: The Struggle against Apartheid.* Ithaca, NY: Cornell University Press, 1999.

Knox, Macgregor, and Williamson Murray, eds. *The Dynamics of Military Revolution 1300–2050.* Cambridge: Cambridge University Press, 2001.

Konaev, Margarita, Husanjot Chahal, Ryan Fedasiuk, Tina Huang, and Ilya Rahkovsky. *U.S. Military Investments in Autonomy and A.I.: A Budgetary Assessment.* Washington, DC: Center for Security and Emerging Technology, 2020. https://cset.georgetown.edu/research/u-s-military-investments-in-autonomy-and-ai-a-budgetary-assessment/.

Krasner, Stephen. "Sovereignty: An Institutional Perspective." *Comparative Political Studies* 21, no. 1 (1988): 66–94.

Krepinevich, Andrew F., Jr. *The Army and Vietnam.* Baltimore, MD: Johns Hopkins Press, 1986.

Krepinevich, Andrew F., Jr. "Keeping Pace with the Military-Technological Revolution." *Issues in Science and Technology* 10, no. 4 (1994): 23–29.

Krepinevich, Andrew F., Jr. *The Military-Technical Revolution: A Preliminary Assessment.* Washington, DC: CSBA, 1991.

Krepinevich, Andrew, and Barry Watts. *The Last Warrior: Andrew Marshall and the Shaping of Modern American Defense Strategy.* New York: Basic Books, 2015.

Kreps, Sarah Elizabeth. *Drones: What Everyone Needs to Know.* Oxford: Oxford University Press, 2016.

Krulak, Victory H. *First to Fight: An Inside View of the U.S. Marine Corps.* Annapolis, MD: Naval Institute Press, 1984.

Kull, Steven, I. M. Destler, and Clay Ramsay. *The Foreign Policy Gap: How Policymakers Misread the Public.* College Park: Center for International and Security Studies at Maryland, 1997.

Kull, Steven, and Clay Ramsay. "Challenging U.S. Policymakers' Image of an Isolationist Public." *International Studies Perspectives* 1, no. 1 (2000): 105–117.

LaFeber, Walter. "The Rise and Fall of Colin Powell and the Powell Doctrine." *Political Science Quarterly* 124, no. 1 (2009): 71–95.

Laird, Melvin R., Secretary of Defense, Testimony before the Committee on Appropriations, House of Representatives, February 25, 1970.

Lake, Daniel. *The Pursuit of Technological Superiority and the Shrinking American Military.* New York: Palgrave, 2019.

Lancaster, Steven L., and Roland P. Hart. "Military Identity and Psychological Functioning: A Pilot Study." *Military Behavioral Health* 3, no. 1 (2015): 83–87.

Larson, Eric Victor. *Casualties and Consensus: The Historical Role of Casualties in Domestic Support for US Military Operations.* Santa Monica, CA: RAND Corporation, 1996.

Lebow, Richard Ned. "Miscalculation in the South Atlantic: The Origins of the Falkland War." *Journal of Strategic Studies* 6, no. 1 (1983): 5–35.

Levie, Howard S. *Mine Warfare at Sea.* New York: Martinus Nijhoff Publishers, 1992.

Lewis, Larry. *Insights for the Third Offset: Addressing Challenges of Autonomy and Artificial Intelligence in Military Operations.* Arlington, VA: Center for Naval Analyses, 2017.

Libicki, Martin C. *The Mesh and the Net: Speculations on Armed Conflict in a Time of Free Silicon.* Washington, DC: National Defense University, March 1994.

Lindsay, James. "Parochialism, Policy, and Constituency Constraints: Congressional Voting on Strategic Weapons Systems." *American Journal of Political Science* 34, no. 4 (1990): 936–960.

Lindsay, Jon. *Information Technology and Military Power.* Ithaca, NY: Cornell University Press, 2020.

Linn, Brian. "The American Way of War Revisited." *Journal of Military History* 66, no. 2 (2002): 501–533.

Linn, Brian. *The Echo of Battle.* Cambridge, MA: Harvard University Press, 2009.

Livingston, Steven, and Todd Eachus. "Humanitarian Crises and US Foreign Policy: Somalia and the CNN Effect Reconsidered." *Political Communication* 12, no. 4 (1995): 413–429.

Lock-Pullan, Richard. "How to Rethink War: Conceptual Innovation and AirLand Battle Doctrine." *Journal of Strategic Studies* 28, no. 4 (2005): 679–702.

Lomov, Nikolai A., ed. *Scientific-Technical Progress and the Revolution in Military Affairs (A Soviet View).* Washington, DC: US Air Force Government Printing Office, 1974.

Long, Jeffrey W. *The Evolution of US Army Doctrine: From Active Defense to AirLand Battle and Beyond.* Fort Leavenworth, KS: US Army Command and General Staff College, 1991.

Lonnquest, John C., and David F. Winkler. *To Defend and Deter: The Legacy of the United States Cold War Missile Program.* Special Report 97/01. Washington, DC: USACERL, 1996.

Ludvigsen, Eric C. "Future Combat Systems: A Status Report." *Army Magazine* 41, no. 2 (February 1991): 38–43.

Luke, Ivan T. "Naval Operations in Peacetime: Not Just 'Warfare Lite.'" *Naval War College Review* 66, no. 2 (2013): Article 4. https://digital-commons.usnwc.edu/nwc-review/vol66/iss2/4.

Macdonald, Julia M. "Eisenhower's Scientists: Policy Entrepreneurs and the Test Ban Debate 1954–1958." *Foreign Policy Analysis* 11, no. 1 (2015): 1–21.

Macdonald, Julia, and Jacquelyn Schneider. "Battlefield Responses to New Technologies Views from the Ground on Unmanned Aircraft." *Security Studies* 28, no. 2 (February 2019) 216–249.

Macdonald, Julia, and Jacquelyn Schneider. "Presidential Risk Orientation and Force Employment Decisions: The Case of Unmanned Weapons." *Journal of Conflict Resolution* 61, no. 3 (2017): 511–536.

MacKenzie, Donald. *Inventing Accuracy: A Historical Sociology of Nuclear Missile Guidance.* Cambridge, MA: MIT Press, 1993.

Maddrell, Debra O. *Quiet Transformation: The Role of the Office of the Net Assessment.* Washington, DC: National War College, May 2, 2003.

Magee, John Gillespie. "High Flight." Accessed December 31, 2023. https://nationalpoetryday.co.uk/poem/high-flight/.

Magnuson, Stew. "Future Combat Systems Didn't Truly Die." *National Defense Magazine*, September 2017. https://www.nationaldefensemagazine.org/articles/2017/9/26/future-combat-systems-didnt-truly-die.

Mahan, Dennis Hart. *A Complete Treatise on Field Fortification, with the General Outlines of the Principles Regulating the Arrangement, the Attack, and the Defense of Permanent Works.* New York: Greenwood Press, 1968.

Mahnken, Thomas G. "Andrew Marshall in Memoriam." War on the Rocks, April 8, 2019. https://warontherocks.com/2019/04/andrew-w-marshall-in-memoriam/.

Mahnken, Thomas. *Technology and the American Way of War since 1945.* New York: Columbia University Press, 2008.

Mahnken, Thomas G., and James R. Fitzsimonds. *The Limits of Transformation: Officer Attitudes Toward the Revolution in Military Affairs.* Newport, RI: Naval War College Press, 2003.

Mahnken, Thomas, and James R. Fitzsimonds. "Revolutionary Ambivalence: Understanding Officer Attitudes toward Transformation." *International Security* 28, no. 2 (2003): 112–148.

Mahoney, James. "Path Dependence in Historical Sociology." *Theory and Society* 29, no. 4 (2000): 507–548.

Mahoney, James, and Celso M. Villegas. "Historical Enquiry and Comparative Politics." In *The Oxford Handbook of Comparative Politics*, edited by Carles Boix and Susan Stokes, 73–89. New York: Oxford University Press, 2007.

Malkasian, Carter. "Toward a Better Understanding of Attrition: The Korean and Vietnam Wars." *Journal of Military History* 68, no. 3 (2004): 911–942.

Mandelbaum, Michael. "Vietnam: The Television War." *Daedalus* 111, no. 4 1982): 157–169.

Mansoor, Peter, and Williamson Murray, eds. *The Culture of Military Organizations*. Cambridge: Cambridge University Press, 2019.

Marcus, Raphael D. "Military Innovation and Tactical Adaptation in the Israel–Hizballah Conflict: The Institutionalization of Lesson-Learning in the IDF." *Journal of Strategic Studies* 38, no. 4 (2015): 500–528.

Marine Corps. "Unmanned Vehicle Unit Changes Name." News release, February 9, 1996.

Martinage, Robert. *Toward a New Offset Strategy: Exploiting U.S. Long Term Advantages to Restore U.S. Global Power Projection Capability*. Washington, DC: Center for Strategic and Budgetary Assessments, 2014.

Mayer, Kenneth. *The Development of the Advanced Medium-Range Air-to-Air Missile: A Case Study of Risk and Reward in Weapon System Acquisition*. Santa Monica, CA: RAND Corporation, 1994.

Mazarr, Michael J. *The Military-Technical Revolution: A Structural Framework*. Washington, DC: Center for Strategic and International Studies, 1993.

McCormick, Michael. "The New FM 100-5: A Return to Operational Art." *Military Review* 78, no. 5 (1997): 3–14.

McCrisken, Trevor. "Obama's Drone War." *Survival: Global Politics and Strategy* 55, no. 2 (2013): 97–122.

McCurley, Thomas. "Molding Generations of Air Force Superiority." Afcent.af.mil, March 20, 2011. https://www.afcent.af.mil/News/Commentaries/Display/Article/223247/molding-generations-of-air-force-superiority/.

McFate, Montgomery. "Being There: US Navy Organizational Culture and the Forward Presence Debate." *Defense and Security Analysis* 36, no. 1 (2021): 42–64.

McGrath, Bryan. "Comments during Hearing before the Subcommittee on Seapower and Projection of Forces." *Unmanned Carrier-Launched Airborne Surveillance and Strike (UCLASS) Requirements Assessment*, July 16, 2014.

McGrath, James R. "Twenty-First Century Information Warfare and the Third Offset Strategy." *Joint Force Quarterly* 82, no. 3 (2016): 16–23.

McMaster, H. R. *Dereliction of Duty*. New York: Harper Collins, 1997.

Menoher, Paul E. "Force XXI: Redesigning the Army through Warfighting Experiments." *Military Intelligence Professional Bulletin*, February 1996. https://fas.org/irp/agency/army/mipb/1996-2/menoher1.htm.

Mermin, Jonathan. "Television News and American Intervention in Somalia: The Myth of a Media-Driven Foreign Policy." *Political Science Quarterly* 112, no. 3 (1997): 385–403.

Metz, Steven, and James Kievit. *Strategy and the Revolution in Military Affairs: From Theory to Policy*. Carlisle, PA: Strategic Studies Institute, 1995.

Middup, Luke. "The Impact of Vietnam on U.S. Strategy in the First Gulf War." *Comparative Strategy* 29, no. 5 (2010): 389–404.

Middup, Luke. *The Powell Doctrine and US Foreign Policy*. New York: Routledge, 2016.

Miller, Laura L., and Charles Moskos. "Humanitarians or Warriors?: Race, Gender, and Combat Status in Operation Restore Hope." *Armed Forces and Society* 21, no. 4 (1995): 615–637.

Millett, Allan R., and Williamson Murray, eds. *Military Effectiveness*. Vol. 2, *The Interwar Period*. Cambridge: Cambridge University Press, 2010.

Mindling, George, and Robert Bolton. *U.S. Air Force Tactical Missiles, 1949–1961*. Morrisville, NC: LuLu.Com, 2008.

Mintrom, Michael. *Policy Entrepreneurs and School Choice*. Washington, DC: Georgetown University Press, 2000.

Mintrom, Michael, and Philippa Norman. "Policy Entrepreneurship and Policy Change." *Policy Studies Journal* 37, no. 4 (2009): 649–667.

Moellering, John H. "The Army Turns Inward." *Military Review* 53, no. 7 (1973): 68–83.

Mohr, Charles. "Cruise Missile Passes Test but Its Critics Score, Too." *New York Times*, July 17, 1983.

Mongilio, Heather. "U.S. Sending Vehicle-Mounted Harpoon Launchers for Ukraine Coastal Defense." USNI News, June 15, 2022. https://news.usni.org/2022/06/15/u-s-sending-vehicle-mounted-harpoon-launchers-for-ukraine-coastal-defense.

Morton, James Saint Claire. *Memoir of an American Fortification*. Washington, DC: William A. Harris, 1959.

Mueller, John. "The Impact of Ideas on Grand Strategy." In *The Domestic Bases of Grand Strategy*, edited by Richard N. Rosencrance and Arthur A. Stein, 48–62. Ithaca, NY: Cornell University Press, 1993.

Mueller, John. *War, Presidents and Public Opinion*. New York: Wiley, 1973.

Murray, Williamson. "Contributions of Military Historians." In *Net Assessment and Military Strategy*, edited by Thomas Mahnken, 139–152. Amherst, MA: Cambria Press, 2020.

Nakasone, Paul M. "A Cyber Force for Persistent Operations." *Joint Force Quarterly* 92, no. 1 (2019): 10–14.

National Defense Authorization Act for Fiscal Year 2001, H.R. 5308. 106th Congress (2000). https://www.congress.gov/106/plaws/publ398/PLAW-106publ398.pdf.

National Defense Authorization Act for Fiscal Year 2020, S. 1790. 116th Congress (2019) https://www.congress.gov/116/bills/s1790/BILLS-116s1790enr.pdf.

National Defense Panel. *Transforming Defense: National Security in the 21st Century*. Arlington, VA: National Defense Panel, December 1997.

National Research Council. *Technology Development for Army Unmanned Ground Vehicle*. Washington, DC: National Academies Press, 2003.

Naval War College. *Tactical Problem IX.: Class of June 1919*. Newport, RI: Naval War College, June 1920.

Nechaev, N. "Voennye sistemy sviazi: tendentsii ikhrazvitiia." *Tekhnika i vooruzheniia*, July 5, 1986.

Nemfakos, Charles, Irv Blickstein, Aine Seitz McCarthy, and Jerry M. Sollinger. *The Perfect Storm: The Goldwater-Nichols Act and Its Effect on Navy Acquisition*. Santa Monica, CA: RAND Corporation, 2010.

Neufield, Jacob. *The Development of Ballistic Missiles in the United States Air Force, 1945–1960*. Washington, DC: Office of Air Force History, United States Air Force, 1990.

Neufeld, Jacob, George M. Watson Jr., and David Chenoweth. *Technology and the Air Force: A Retrospective Assessment*. Washington, DC: Air Force History and Museums Program, 2007.

Newcome, Laurence R. *Unmanned Aviation: A Brief History of Unmanned Aerial Vehicles*. Reston, VA: American Institute of Aeronautics and Astronautics, 2004.

Newpower, Anthony. *Iron Men and Tin Fish: The Race to Build a Better Torpedo during World War II*. Westport, CT: Praeger, 2006.

"Night Attack on Ukraine Cost Russia at Least US $120 Million." May 16, 2023. https://www.pravda.com.ua/eng/news/2023/05/16/7402408/.

Nye, Joseph S., and William A. Owens. "America's Information Edge." *Foreign Affairs* 75 (March–April 1996): 23–25.

O'Hanlon, Michael. *Technological Change and the Future of Warfare*. Washington, DC: Brookings Institute, 2000.

"Oral History: Norman Schwarzkopf." *Frontline*. Accessed December 31, 2023. https://www.pbs.org/wgbh/pages/frontline/gulf/oral/schwarzkopf/1.html.

O'Rourke, Ronald. "Defense Primer: Department of the Navy." Federation of American Scientists, August 21, 2020. https://fas.org/sgp/crs/natsec/IF10484.pdf.

O'Rourke, Ronald. *Navy Large Unmanned Surface and Undersea Vehicle: Background and Issues for Congress*. Washington, DC: Library of Congress, Congressional Research Service, 2020. https://fas.org/sgp/crs/weapons/R45757.pdf.

Office of Technology Assessment. *New Technology for NATO: Implementing a Follow-on Forces Attack*. Washington, DC: Government Printing Office, 1987.

Ogarkov, Nikolei. "Sovetskaia voennaia nauka." *Vsegda v gotovnosti*, February 18, 1978.

Olsen, John Andreas. *John Warden and the Renaissance of American Air Power*. Washington, DC: Potomac Books, 1997.

Owens, William A., and Ed Offley. *Lifting the Fog of War*. Baltimore, MD: Johns Hopkins University Press, 2001.

Pannell, Ian, and Allie Weintraub. "Inside Ukraine's Efforts to Bring an 'Army of Drones' to War against Russia." *ABC News*, September 15, 2023. https://abcnews.go.com/US/inside-ukraines-efforts-bring-army-drones-war-russia/story?id=103152130.

Parker, Geoffrey. *The Military Revolution: Military Innovation and the Rise of the West, 1500–1800*. Cambridge: Cambridge University Press, 1996.

Parrott, David. *The Business of War: Military Enterprise and Military Revolution in Early Modern Europe*. Cambridge: Cambridge University Press, 2012.

Peebles, Curtis. *High Frontier: The United States Air Force and the Military Space Program*. Darby, PA: Diane Publishing, 1997.

"Pentagon's Plan for a 500-Ship Navy Calls for More Unmanned Vessels." Maritime-Executive.com, October 7, 2020. https://maritime-executive.com/article/pentagon-s-plan-for-a-500-ship-navy-calls-for-more-unmanned-vessels.

Pernin, Christopher G., Elliot Axelband, Jeffrey A. Drezner, Brian Barber Dille, John Gordon IV, Bruce Held, K. Scott McMahon, Walter L. Perry, Christopher Rizzi, Akhil R. Shah, Peter A. Wilson, and Jerry M. Sollinger. *Lessons from the Army's Future Combat Systems Program*. Santa Monica, CA: RAND Corporation, 2012.

Perry, Robert L. "Appendix." In *Science, Technology, and Warfare: The Proceedings of the 3rd Military History Symposium*, edited by Monte Wright and Lawrence Paszek, 110–122. Colorado Springs: United States Air Force Academy, 1969.

Petraeus, David H. "Lessons of History and Lessons of Vietnam." *Parameters* 40, no. 4 (2010): 48–61.

Pettyjohn, Stacie, and Hannah Dennis. "Precision and Posture: Defense Spending Trends and the FY23 Budget Request." Center for a New American Security, November 17, 2022. https://www.cnas.org/publications/reports/precision-and-posture-defense-spending-tre.

Pfaltzgraff, Robert L., Jr., and Jacquelyn K. Davis. *The Cruise Missile: Bargaining Chip or Defense Bargain*. Cambridge, MA: Institute for Foreign Policy Analysis, 1977.

Polmar, Norman, and John O'Connell. *Strike from the Sea: The Development and Deployment of Strategic Cruise Missiles since 1934*. Annapolis, MD: Naval Institute Press, 2020.

Pomerlau, Mark. "Cyber Command Prepares to Gain Significant Budget Control." Fedscoop, March 4, 2022. https://www.fedscoop.com/cyber-command-budget-control-preparations-pom.

Ponce, B. P. *Hollow Promises: The Problem of Culture and the Integration of New Technology into the Navy*. Fort Leavenworth, KS: Army Command and General Staff College, 2004.

Popkov, M.G. "Metodologicheskii analiz informatsionnykh protsessov v sisteme'chelovek-voennaia tekhnika." PhD diss., Voenno-Politicheskaia Akademiia, 1983.

Posen, Barry. *The Sources of Military Doctrine: France, Britain, and Germany between the World Wars*. Ithaca, NY: Cornell University Press, 1986.

Powell, Colin. *My American Journey*. New York: Ballantine Books, 1996.

Powell, Colin. "U.S. Forces: Challenges Ahead." *Foreign Affairs*, Winter 1992/1993. https://www.cfr.org/world/us-forces-challenges-ahead/p7508.

Power, Maj. Gen. Thomas. "Letter to Director of Requirements." HQ/USAF, March 30, 1953.

"Pressure for New Bomber Rises in Congress." *Aviation Week and Space Technology*, February 11, 1980.

Price, Richard. "Reversing the Gun Sights: Transnational Civil Society Targets Land Mines." *International Organization* 52, no. 3 (1998): 613–644.

Probasco, Emelia. "The Future of Drones in Ukraine." CSET, November 13, 2023. https://cset.georgetown.edu/article/the-future-of-drones-in-ukraine-a-report-from-the-diu-brave1-warsaw-conference/.

Rabe, Barry. *Statehouse and Greenhouse: The Stealth Politics of America Climate Change Policy.* Washington, DC: Brookings Institution Press, 2004.

Randolph, Stephen P. *Powerful and Brutal Weapons: Nixon, Kissinger, and the Easter Offensive.* Cambridge, MA: Harvard University Press, 2009.

Rearick, Perry D. "Force Protection and Mission Accomplishment in Bosnia and Herzegovina." Thesis, US Army Command and General Staff College, 2001.

Record, Jeffrey. "Force-Protection Fetishism: Sources, Consequences, and Solutions." *Air Space Power Journal* 14, no. 2 (Summer 2000): 4–11.

Reiter, Dan, and Allan C. Stam. *Democracies at War.* Princeton, NJ: Princeton University Press, 2002.

Rems, Alan. "A Propaganda Machine Like Stalin's." *Naval History Magazine* 33, no. 3 (2019). https://www.usni.org/magazines/naval-history-magazine/2019/june/propaganda-machine-stalins.

Renshon, Jonathan. "Stability and Change in Belief Systems: The Operational Code of George W. Bush." *Journal of Conflict Resolution* 52, no. 6 (2008): 814–836.

Ricks, Thomas E. *Fiasco: The American Military Adventure in Iraq.* London: Penguin UK Books, 2006.

Rid, Thomas. *Rise of the Machines: A Cybernetic History.* New York: W. W. Norton & Company, 2016.

Risse-Kappen, Thomas, Stephen C. Ropp, and Kathryn Sikkink, eds. *The Power of Human Rights: International Norms and Domestic Change.* New York: Cambridge University Press, 1999.

Robinson, Piers. *The CNN Effect: The Myth of News, Foreign Policy and Intervention.* New York: Routledge, 2002.

Robinson, Todd C. "What Do We Mean by Nuclear Proliferation." *Nonproliferation Review* 22, no. 1 (2015): 53–70.

Rogers, Clifford, ed. *The Military Revolution Debate: Readings on the Military Transformation of Early Modern Europe.* Boulder, CO: Westview Press, 1995.

Romjue, John L. *From Active Defense to Airland Battle: The Development of Army Doctrine, 1973–1982.* Fort Monroe, VA: Historical Office, US Army Training and Doctrine Command, 1984.

Romjue, John L. *A History of Army 86.* Vol. 2, *The Development of the Light Division, the Corps, and Echelons above Corps November 1979–December 1980.* Fort Monroe, VA: Historical Office, US Army Training and Doctrine Command, 1982.

Rosen, Stephen P. "The Impact of the Office of Net Assessment on the American Military in the Matter of the Revolution in Military Affairs," *Journal of Strategic Studies* 34, no. 4 (2010): 469–482.

Rosen, Stephen P. *Winning the Next War: Innovation and the Modern Military.* Ithaca, NY: Cornell University Press, 1994.

Rothkopf, David J. *Running the World: The Inside Story of the National Security Council and the Architects of American Power.* New York: PublicAffairs, 2005.

Rundquist, Barry, and Thomas M. Carsey. *Congress and Defense Spending: The Distributive Politics of Military Procurement.* Norman: University of Oklahoma Press, 2002.

Russett, Bruce. "Doves, Hawks, and US Public Opinion." *Political Science Quarterly* 105, no. 4 (1990): 515–538.

Russett, Bruce. "A More Democratic and Therefore a More Peaceful World." *Journal of New Paradigm Research* 29, no. 4 (1990): 243–263.

Sapolsky, Harvey, Eugene Gholz, and Caitlin Talmadge. *US Defense Politics: The Origins of Security Policy.* New York: Taylor & Francis, 2017.

Sarkasian, Sam. "U.S. National Security Strategy: The Next Decade." In *The U.S. Army in a New Security Era,* edited by Sam Sarkasian and John Allen Williams, 1–18. Boulder, CO: Lynne Reiner Publishers, 1990.

Saunders, Elizabeth. *Leaders at War: How Presidents Shape Military Interventions.* Ithaca, NY: Cornell University Press, 2011.

Saunders, Elizabeth N. "No Substitute for Experience: Presidents, Advisers, and Information in Group Decision Making." *International Organization* 71 (Supplement 2017): S219–S247.

Scales, Bob. *Certain Victory: The US Army in the Gulf War.* Washington, DC: Potomac Books, 1997.

Scharre, Paul. *Army of None: Autonomous Weapons and the Future of War.* New York: W. W. Norton & Company, 2018.

Schimmelfennig, Frank. "The Community Trap: Liberal Norms, Rhetorical Action, and the Eastern Enlargement of the European Union." *International Organization* 55, no. 1 (2001): 47–80.

Schneider, Jacquelyn. "The Information Revolution and International Stability: A Multi-Article Exploration of Computing, Cyber, and Incentives for Conflict." PhD diss., George Washington University, 2017.

Schneider, Jacquelyn. "Persistent Engagement: Foundation, Evolution, and Evaluation of a Strategy." Lawfare, May 10, 2019. https://www.lawfareblog.com/persistent-engagement-foundation-evolution-and-evaluation-strategy.

Schneider, Jacquelyn. "What It Will Take to Supply Ukraine for the Long Haul." *Wall Street Journal,* July 7, 2022. https://www.wsj.com/articles/what-it-will-take-to-supply-ukraine-for-the-long-haul-11657206373.

Schneider, Jacquelyn G., Emily O. Goldman, Michael Warner, Paul M. Nakasone, Chris C. Demchak, Nancy A. Norton, Joshua Rovner, et al. *Ten Years In: Implementing Strategic Approaches to Cyberspace.* Newport, RI: US Naval College War Press, 2020.

Schneider, Jacquelyn, and Julia Macdonald. "Looking Back to Look Forward: Autonomous Systems, Military Revolutions, and the Importance of Cost." *Journal of Strategic Studies* 47, no. 2 (2023): 162–184.

Schneider, Jacquelyn, and Julia Macdonald. "U.S. Public Support for Drone Strikes: When Do Americans Prefer Unmanned over Manned Platforms?" Center for a New American Security, September 2016. https://www.cnas.org/publications/reports/u-s-public-support-for-drone-strikes.

Schultz, Timothy. *The Problem with Pilots.* Baltimore, MD: Johns Hopkins Press, 2018.

Schwartz, Norman, and Suzie Schwartz, with Ron Levinson. *Journey: Memoirs of an Air Force Chief of Staff.* New York: Skyhorse Publishing, 2018.

Schwarz, Benjamin C. *Casualties, Public Opinion, and US Military Intervention: Implications for US Regional Deterrence Strategies.* Santa Monica, CA: RAND Corporation, 1994.

Schwarzkopf, Norman H., with Peter Petre. *The Autobiography: It Doesn't Take a Hero.* London: Bantam Books, 1993.

Sims, John. *Shackled by Perceptions: America's Desire for Bloodless Interventions.* Maxwell AFB, AL: School of Advanced Air Power Studies, 1997.

Singer, Peter W. "The Future of War Will be Robotic." CNN Opinion, last modified February 23, 2015. https://www.cnn.com/2015/02/23/opinion/singer-future-of-war-robotic/index.html.

Singh, Sonali, and Christopher R. Way. "The Correlates of Nuclear Proliferation: A Quantitative Test." *Journal of Conflict Resolution* 48, no. 6 (December 2004): 859–885.

Siverson, Randolph M. "Democracies and War Participation: In Defense of the Institutional Constraints Argument." *European Journal of International Relations* 1, no. 4 (1995): 481–489.

Skrentny, John David. *The Ironies of Affirmative Action: Politics, Culture, and Justice in America.* Chicago: University of Chicago Press, 2018.

Slayton, Rebecca. *Arguments That Count: Physics, Computing, and Missile Defense.* Cambridge, MA: MIT Press, 2013.

Smit, Wim, and John Grin. *Military Technological Innovation and Stability in a Changing World: Politically Assessing and Influencing Weapon Innovation and Military Research and Development.* Charlotte: Virginia University Press, 1992.

Smith, Rogers. "Identities, Interests, and the Future of Political Science." *Perspectives on Politics* 2, no. 2 (2004): 301–312.

Snider, Don M., John A. Nagl, and Tony Pfaff. *Army Professionalism, the Military Ethic, and Officership in the 21st Century.* Carlisle, PA: Army War College, Strategic Studies Institute, 1999.

Solyanking, A. G., M. V. Pavlov, and I. V. Pavlov. *Domestic Armored Vehicles, Twentieth Century: 1905–1941.* Moscow: Elksprint, 2002.

Spaeder, Leo. "Sir, Who Am I? An Open Letter to the Commandant of the Marines." War on the Rocks, March 28, 2019. https://warontherocks.com/2019/03/sir-who-am-i-an-open-letter-to-the-incoming-commandant-of-the-marine-corps/.

Sterling, Christopher H. *Military Communications: From Ancient Times to the 21st Century.* Santa Barbara, CA: ABC-CLIO, 2008.

Stevenson, Charles A. "The Evolving Clinton Doctrine on the Use of Force." *Armed Force Society* 22, no. 4 (1996): 511–535.

Stine, G. Harry. *ICBM: The Making of the Weapon that Changed the World.* New York: Orion Books, 1991.

Strobel, Warren P. "The CNN Effect." *American Journalism Review* 18, no. 4 (1996): 32–38.

Stumpf, David K. *Minuteman: A Technical History of the Missile That Defined American Nuclear Warfare.* Fayetteville: University of Arkansas Press, 2021.

Sullivan, Gen. Gordon R., and Lt. Col. Anthony M. Coroalles. *The Army in the Information Age.* Carlisle, PA: US Army War College Strategic Studies Institute, March 1995.

Sullivan, Gen. Gordon R., and Lt. Col. James M. Dubik. *Land Warfare in the 21st Century.* Carlisle, PA: US Army War College Strategic Studies Institute, February 1993.

Survey by Louis Harris and Associates, "Harris 1966 Survey No. 1522." In Harris Collection, Odum Institute, Louis Harris Data Center, University of North Carolina. https://hdl.handle.net/1902.29/H-1522

Survey by Louis Harris and Associates, "Harris 1966 Survey No. 1531." In Harris Collection, Odum Institute, Louis Harris Data Center, University of North Carolina. https://hdl.handle.net/1902.29/H-1531.

Sustein, Cass. *Free Markets and Social Justice.* New York: Oxford University Press, 1997.

"The System of Power and Democratic Institutions." *Social Sciences (Moscow)* 6, no. 3 (1975): 122.

Tannenwald, Nina. *The Nuclear Taboo.* Cambridge: Cambridge University Press, 2007.

Thielmann, Greg, "Looking Back: The Missile Gap Myth and its Progeny." *Arms Control Today* 41, no. 4 (2011): 44–48.

Toffler, Alvin, and Heidi Toffler. *Creating a New Civilization: The Politics of the Third Wave.* Atlanta, GA: Turner Publishing, 1995.

Toffler, Alvin, and Heidi Toffler. "Getting Set for the Coming Millennium." *The Futurist* 29, no. 2 (1995): 10–15.

Toffler, Alvin, and Heidi Toffler. *War and Anti-War Survival at the Dawn of the 21st Century* Boston: Little, Brown, 1993.

Tomes, Robert. *US Defense Strategy from Vietnam to Operation Iraqi Freedom*. New York: Routledge, 2007.

Tomz, Michael. "Domestic Audience Costs in International Relations: An Experimental Approach." *International Organization* 61, no. 4 (2007): 821–840.

Tondo, Lorenzo. "Sea Mines: The Deadly Danger Lurking in Ukraine's Waters." *The Guardian*, July 11, 2022. https://www.theguardian.com/world/2022/jul/11/sea-mines-ukraine-waters-russia-war-black-sea.

Trevithick, Joseph. "The US Navy Has Created Its First Ever Underwater Drone Squadron." TheDrive.com, September 28, 2017. https://www.thedrive.com/the-war-zone/14733/the-us-navy-has-created-its-first-ever-underwater-drone-squadron.

Truver, Scott C. "Naval Mines and Mining: Innovating in the Face of Benign Neglect." CIMSEC.org, December 20, 2016. https://cimsec.org/naval-mines-mining-innovating-face-benign-neglect/.

US Army Training and Doctrine Command. *TRADOC Pamphlet 525-5: Force XXI Operations*. Fort Monroe, VA: US Army TRADOC, August 1 1994.

US Department of Defense. *Conduct of the Persian Gulf Conflict: An Interim Report to Congress*. July 1991.

US Department of Defense. *Joint Vision 2020: America's Military: Preparing for Tomorrow*. 2000. https://www.hsdl.org/?abstract&did=446826.

US Department of Defense. *Quadrennial Defense Review Report*. 2001. https://archive.defense.gov/pubs/qdr2001.pdf.

US Department of Defense. *Unmanned Systems Roadmap 2007–2032*. Washington, DC: Office of the Secretary of Defense, 2007. https://www.hsdl.org/?abstract&did=481851.

US Department of State. *The Reagan Administration and Lebanon, 1981–1984*. Washington, DC: Office of the Historian, Foreign Service Institute, 1985. https://history.state.gov/milestones/1981-1988/lebanon.

Valavanis, Kimon P., ed. *Advances in Unmanned Aerial Vehicles*. Dordrecht: Springer, 2007.

Valavanis, Kimon P., and George J. Vachtsevanos, eds. *Handbook of Unmanned Aerial Vehicles*. Dordrecht: Springer, 2015.

Vickers, Michael G. *Warfare in 2020: A Primer*. Washington, DC: Center for Strategic and Budgetary Assessments, 1996.

Voeten, Erik, and Paul R. Brewer. "Public Opinion, the War in Iraq, and Presidential Accountability." *Journal of Conflict Resolution* 50, no. 6 (2006): 809–830.

Volkov, G. N. *Man and the Challenge of Technology*. Moscow: Novosti, 1972.

Waddell, Timothy. "Marshal N.V. Ogarkov and the Transformation in Military Soviet Affairs." Thesis, University of Manitoba, 1999.

Walton, Timothy. "Securing the Third Offset Strategy." *Joint Force Quarterly* 82 (2016): 6–15.

Waltzer, Michael. "Targeted Killing and Drone Warfare." *Dissent*, January 11, 2013.

Warden, John A. *The Air Campaign: Planning for Combat*. Darby, PA: DIANE Publishing, 1998.

Weber, Max. *Essays in Sociology*, 267–301. New York: Oxford University Press, 1946.

Weigley, Russell Frank. *The American Way of War: A History of United States Military Strategy and Policy*. Bloomington: Indiana University Press, 1977.

"The Weinberger Doctrine." *Washington Post*, November 30, 1984. https://www.washingtonpost.com/archive/politics/1984/11/30/the-weinberger-doctrine/c7f20ffe-b591-4189-ad05-a704aac1935d/.

Wells, Anthony R. *A Tale of Two Navies*. Annapolis, MD: Naval Institute Press, 2017.

Wells, Mark. *Tribal Warfare: The Society of Modern Airmen.* Maxwell AFB, AL: Air University, 2015.

Werrell, Kenneth. *Chasing the Silver Bullet: US Air Force Weapons Development from Vietnam to Desert Storm.* Washington, DC: Smithsonian Institute Press, 2003.

Werrell, Kenneth. *The Evolution of the Cruise Missile.* Maxwell, AL: Air University Press, 1985.

Westmoreland, William. *A Soldier Reports.* New York: Doubleday, 1976.

Whitmore, Bishane A. *Evolution of Unmanned Aerial Warfare: A Historical Look at Remote Airpower–A Case Study in Innovation.* Fort Leavenworth, KS: US Army Command and General Staff College, 2016.

Whittle, Richard. *Predator: The Secret Origins of the Drone Revolution.* New York: Henry Holt and Company, 2014.

Wildenberg, Thomas, and Norman Polmar. *Ship Killers: A History of the American Torpedo.* Annapolis, MA: Naval Institute Press, 2010.

Wilson, George C. "Senate Chairman Pushes Unmanned Warfare." *Government Executive,* March 6, 2000. https://www.govexec.com/federal-news/2000/03/senate-chairman-pushes-unmanned-warfare/1977/.

Wilson, George C. "Unmanned Gain Backing." *Washington Post,* September 25, 1985. https://www.washingtonpost.com/archive/politics/1985/09/05/unmanned-weapons-gain-backing/00fd9af9-da42-42c6-b174-6f9f50a08e26/.

Woods, Kevin S. *Limiting Casualties: Imperative or Constraint?* Fort Leavenworth, KS: Army Command and General Staff College, School of Advanced Military Studies, 1997.

Woodward, Bob. *The Commanders.* New York: Touchstone Books, 1991.

Yenne, Bill. *Attack of the Drones: A History of Unmanned Aerial Combat.* Saint Paul, MN: MBI Publishing Company, 2004.

Young, Michael James. "The US Air Force's Long Range Detection Program and Project MOGUL." *Air Power History* 67, no. 4 (2020): 25–32.

Youngblood, Norman. *The Development of Mine Warfare: A Most Murderous and Barbarous Conduct.* Westport, CT: Praeger Security International, 2006.

Youssef, Nancy A., and Doug Cameron. "U.S. Expected to Send Ukraine Longer-Range Smart Bombs in Next Aid Package." *Wall Street Journal,* February 1, 2023. https://www.wsj.com/articles/u-s-expected-to-send-ukraine-longer-range-smart-bombs-in-next-aid-package-11675270504.

Zaloga, Steven J. *Unmanned Aerial Vehicles: Robotic Air Warfare 1917–2007.* London: Bloomsbury Publishing, 2011.

Zaloga, Steven. *V-1 Flying Bomb 1942–52: Hitler's Infamous "Doodlebug."* London: Bloomsbury Publishing, 2011.

Zegart, Amy. "Cheap Fights, Credible Threats: The Future of Armed Drones and Coercion." *Journal of Strategic Studies* 43, no. 1 (2020): 1–41.

Ziegler, Charles A. "Weapons Development in Context: The Case of the World War I Balloon Bomber." *Technology and Culture* 35, no. 4 (1994): 750–767.

Zimmerman, S. Rebecca, Kimberley Jackson, Natasha Lander, Colin Roberts, Dan Madden, and Rebeca Orrie. *Movement and Maneuver: Culture and the Competition for Influence among the U.S. Military Services.* Santa Monica, CA: RAND Corporation, 2019.

Zinni, Anthony. "A Conversation with General Anthony Zinni." Interview by Bruce R. Kuniholm and Alex Roland, Duke University Living History Program, August 23, 2016. http://livinghistory.sanford.duke.edu/interviews/anthony-zinni/.

Index

For the benefit of digital users, indexed terms that span two pages (e.g., 52–53) may, on occasion, appear on only one of those pages.

Tables are indicated by an italic *t* following the page number

2 04